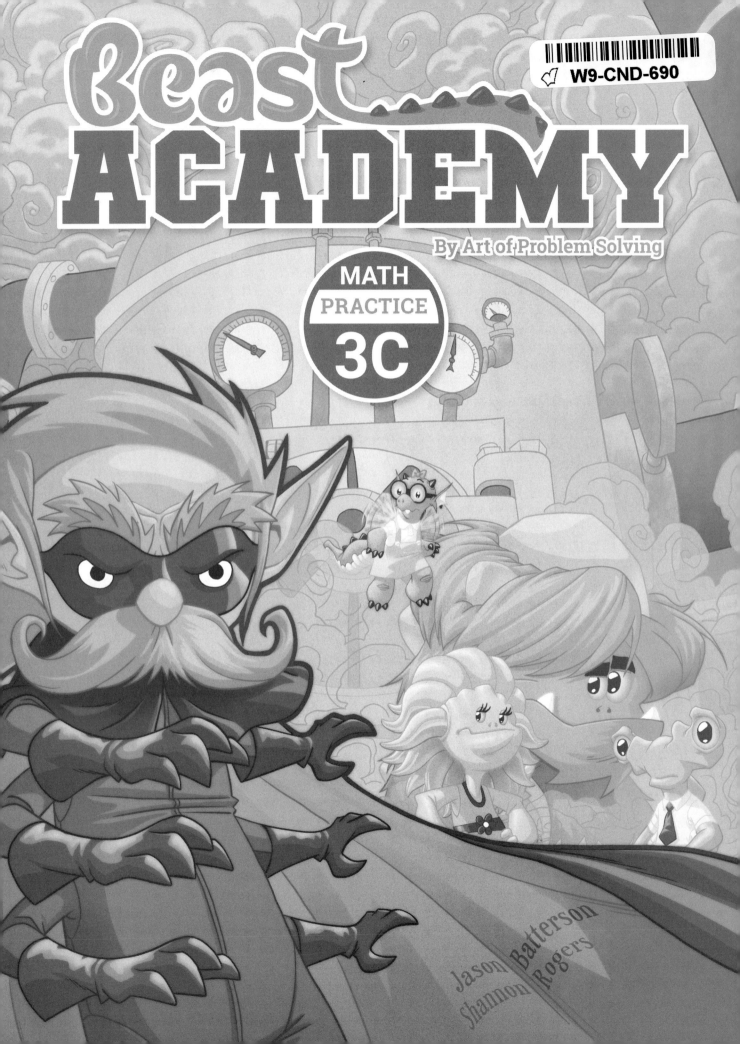

Published by: AoPS Incorporated
                     10865 Rancho Bernardo Rd Ste 100
                     San Diego, CA 92127-2102
                     info@BeastAcademy.com

ISBN: 978-1-934124-45-1

Written by Jason Batterson and Shannon Rogers
Book Design by Lisa T. Phan
Illustrations by Erich Owen
Grayscales by Greta Selman

Visit the Beast Academy website at BeastAcademy.com.
Visit the Art of Problem Solving website at artofproblemsolving.com.
Printed in the United States of America.
2020 Printing.

Contents:

This is Practice Book 3C in the Beast Academy level 3 series.

3A
· Shapes
· Skip-Counting
· Perimeter and Area

3B
· Multiplication
· Perfect Squares
· The Distributive Property

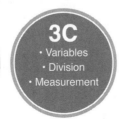
3C
· Variables
· Division
· Measurement

3D
· Fractions
· Estimation
· Area

For more resources and information, visit BeastAcademy.com.

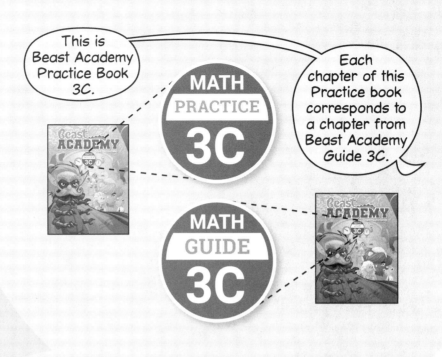

This is Beast Academy Practice Book 3C.

Each chapter of this Practice book corresponds to a chapter from Beast Academy Guide 3C.

MATH PRACTICE 3C

MATH GUIDE 3C

The first page of each chapter includes a recommended sequence for the Guide and Practice book.

You may also read the entire chapter in the Guide before beginning the Practice chapter.

Use this Practice book with Guide 3C from BeastAcademy.com.

Recommended Sequence:

Book	Pages
Guide:	12–23
Practice:	7–11
Guide:	24–28
Practice:	12–21
Guide:	29–41
Practice:	22–34

You may also read the entire chapter in the Guide before beginning the Practice chapter.

Some problems in this book are very challenging. These problems are marked with a ★. The hardest problems have two stars!

Every problem marked with a ★ has a *hint!*

Hints for the starred problems begin on page 100.

Other problems are marked with a ✏. For these problems, you should write an explanation for your answer.

54.
★

55.
✏

42   Guide Pages: 39-43

Some pages direct you to related pages from the Guide.

None of the problems in this book require the use of a calculator.

Solutions are in the back, starting on page 104.

A complete explanation is given for every problem!

CHAPTER 7
Variables

Use this Practice book with
Guide 3C from BeastAcademy.com.

Recommended Sequence:

Book	Pages:
Guide:	12–23
Practice:	7–11
Guide:	24–28
Practice:	12–21
Guide:	29–41
Practice:	22–34

You may also read the entire chapter
in the Guide before beginning the
Practice chapter.

A **variable** is a symbol that is used to represent a quantity.

It's a letter or other symbol that stands for a number!

EXAMPLE | What number does the ☐ represent in the equation below?

40+☐=48.

40+8=48.

So, ☐ represents **8**.

PRACTICE | What number does the ☐ represent in each equation below?

1. 9+☐=16.

1. _____

2. 35−☐=20.

2. _____

3. 80+20=☐.

3. _____

4. ☐−9=31.

4. _____

PRACTICE | What number does n represent in each equation below?

5. 30+n=130.

5. $n=$_____

6. n+5=35.

6. $n=$_____

7. n−3=67.

7. $n=$_____

8. 100−n=40.

8. $n=$_____

When we **evaluate** an expression that includes one or more variables, we replace each variable with a number.

Then, we find the value of the expression. Remember to use the correct order of operations!

EXAMPLE | Evaluate the expression below when $m=5$.

$$24-m\times2$$

Since $m=5$, we replace the m in $24-m\times2$ with 5:

$$24-5\times2.$$

Then, we evaluate $24-5\times2$:

$$24-5\times2=24-10=\textbf{14}.$$

PRACTICE | Evaluate each expression below when $n=6$.

9. $13+n$

9. _____

10. $33-n$

10. _____

11. $n\times4$

11. _____

12. $2\times4+n$

12. _____

PRACTICE | Evaluate each expression below when $r=10$.

13. $r+216$

13. _____

14. $152+9-r$

14. _____

15. $122\times r$

15. _____

16. $7\times(r+20)$

16. _____

EXAMPLE | Evaluate 16−*t* when *t*=4.

When *t*=4, the expression 16−*t* is equal to 16−4=**12**.

PRACTICE | Evaluate 9×*a*+3 for each value of *a*.

17. *a*=9

17._____

18. *a*=4

18._____

19. *a*=20

19._____

PRACTICE | Evaluate 300−*k*×2 for each value of *k*.

20. *k*=25

20._____

21. *k*=100

21._____

22. *k*=60

22._____

PRACTICE | Evaluate 3×(*d*+4) for each value of *d*.

23. *d*=6

23._____

24. *d*=10

24._____

25. *d*=17

25._____

When you **simplify** an expression, you write it in a way that means the same thing but is easier to use.

EXAMPLE | Simplify the expression below.

$$y+20-6$$

Adding 20 then subtracting 6 is the same as adding 20−6=14. So, we have

$$y+20-6=y+14.$$

We cannot simplify this expression further.

Our simplified expression is **y+14**.

PRACTICE | Simplify each expression below.

26. We know that 9+9+9+9=4×9, and 4+4+4+4+4+4+4=7×4. Simplify $n+n+n+n+n+n$.

26. _____

27. We know that 7−7=0 and 11−11=0. Simplify $n-n$.

27. _____

28. We know that 6+5−5=6 and 9+3−3=9. Simplify $n+5-5$.

28. _____

29. We know that 7+4−4=7 and 8+2−2=8. Simplify $17+n-n$.

29. _____

PRACTICE | Simplify each expression below.

30. $d+11+14$

30. _____

31. $17+p+2$

31. _____

32. $13-12+f$

32. _____

33. $5+k-k$

33. _____

34. $j-20+20$

34. _____

35. $w+20-w$

35. _____

36. ★ $13+g-13-g$

36. _____

37. ★ $t+t-t+t+t-t-t$

37. _____

Expressions can be used to describe patterns!

Grogg makes patterns from toothpicks. He records the numbers of shapes and toothpicks in a table.

EXAMPLE Complete the table below with the number of toothpicks needed to make each number of triangles.

Triangles	Toothpicks
1	3
2	6
3	9
4	
5	
6	
7	
8	
n	

Triangles	Toothpicks
1	3
2	6
3	9
4	**12**
5	**15**
6	**18**
7	**21**
8	**24**
n	**$n\times3$**

For each triangle in the diagram, Grogg needs 3 toothpicks. So, we can multiply the number of triangles by 3 to get the number of toothpicks.

4×3=12.
5×3=15.
6×3=18.
7×3=21.
8×3=24.

← To make a pattern with n triangles, he needs $n\times3$ toothpicks.

PRACTICE | Use the toothpick patterns below for Problems 38-40.

38. Complete the table below with the number of toothpicks needed to make each number of squares.

Squares	Toothpicks
1	4
2	8
3	12
4	
5	
10	
100	

39. Circle the expression below that describes the number of toothpicks that Grogg needs to make a diagram of n squares.

n	

$n+3$ $\qquad$ $n\times3$ $\qquad$ $n\times4$ $\qquad$ $1+n\times2$ $\qquad$ $1+n\times3$

Check your work! Evaluate the expression you chose for some values of n from the table. Your answers should match entries in the table above.

40. How many toothpicks does Grogg need to make the same pattern with 40 squares?

40. _____

PRACTICE | Use the toothpick pattern below for Problems 41-43.

41. Complete the table below with the number of toothpicks needed to make each number of squares.

Squares	Toothpicks
1	4
2	7
3	10
4	
5	
★ 20	
★ 100	

42. Circle the expression below that describes the number of toothpicks that Grogg needs to make a diagram of n squares.

n	

$n×5$ $n×3$ $n+3$ $1+n×2$ $1+n×3$

Check your work! Evaluate the expression you chose for some values of n from the table. Your answers should match entries in the table above.

43. How many toothpicks does Grogg need to make the same pattern with 50 squares?

43. _____

PRACTICE | Use the toothpick pattern below for Problem 44.

44. Complete the table below with the number of toothpicks needed to make each number of triangles.

Triangles	Toothpicks
1	3
2	5
3	7
4	
5	
n	
100	

Make sure to check the expression you write for some values of n!

PRACTICE | Use the toothpick pattern below for Problem 45.

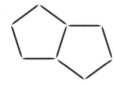 

45. Complete the table below with the number of toothpicks needed to make each number of pentagons.

Pentagons	Toothpicks
1	5
2	9
3	13
4	
5	
n	
100	

EXAMPLE | Write an expression for the perimeter of an equilateral triangle with sides of length t.

Since the triangle is equilateral, we know the length of each side is t.

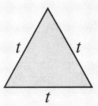

So, the perimeter of the triangle is
$t+t+t$ or $3 \times t$.

PRACTICE

46. Write an expression for the perimeter of a square with side length s.

46. _____

47. Write an expression for the perimeter of a regular hexagon with side length h.

47. _____

48. Write an expression for the **area** of a square with side length s.

48. _____

49. Write an expression for the area of a rectangle with height 5 and width w.

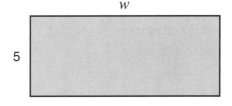

50. Write an expression for the area of a rectangle with height h and width w.

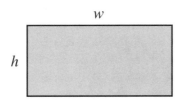

50. _____

51. Write an expression for the perimeter of a rectangle with height h and width 20.

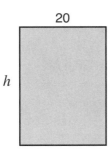

51. _____

52. Write an expression for the perimeter of a rectangle with height h and width w.

52. _____

Calamitous Clod has encoded a riddle below!

Find the value of each variable to unravel my riddle!

PRACTICE | Solve each equation. Then, fill in the blanks with the correct letter to reveal the riddle. You might not use every letter.

53. $2×a=18$ $a=$_____

54. $6+b=12$ $b=$_____

55. $d=2+3×2$ $d=$_____

56. $e×9=45$ $e=$_____

57. $27=9×h$ $h=$_____

58. $40=57−m$ $m=$_____

59. $80=93−o$ $o=$_____

60. $4×r=28$ $r=$_____

61. $70=s×7$ $s=$_____

62. $7=t−15$ $t=$_____

63. $2×10−w=19$ $w=$_____

64. $68+z=79$ $z=$_____

$$\overline{}_{1}\ \overline{}_{3}\ \overline{}_{5}\ \overline{}_{7}\ \overline{}_{5}\qquad \overline{}_{8}\ \overline{}_{13}\qquad \overline{}_{17}\ \overline{}_{9}\ \overline{}_{22}\ \overline{}_{3}$$

$$\overline{}_{6}\ \overline{}_{5}\ \overline{}_{9}\ \overline{}_{10}\ \overline{}_{22}\ \overline{}_{10}\qquad \overline{}_{5}\ \overline{}_{9}\ \overline{}_{22}\ ?$$

Calamitous Clod has encoded the answer to his riddle below.
Be careful! He used a different code for the letters on this page.

PRACTICE | Solve each equation. Then, fill in the blanks with the correct letter to reveal the answer to the riddle. You might not use every letter.

65. $90-a=79$ $a=$_____   **66.** $5×5-2=b$ $b=$_____

67. $d+3=20$ $d=$_____   **68.** $36-e=24$ $e=$_____

69. $g-12=4$ $g=$_____   **70.** $150-i=130$ $i=$_____

71. $5×j=20$ $j=$_____   **72.** $11×l=22$ $l=$_____

73. $100=m×20$ $m=$_____   **74.** $n+50=75$ $n=$_____

75. $2×9+6=s$ $s=$_____   **76.** $t+4=22$ $t=$_____

___ ___   ___   ___ ___ ___ ___ ___
11 18   11   18 20 5 12 24

___ ___ ___ ___ ___ !
18 11 23 2 12

*In the diagrams below, the number in a region is the **sum** of the numbers in the hexagons that surround it.*

EXAMPLE | Find the value of *n* in the diagram shown.

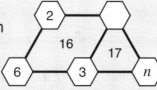

First, we use the region with a sum of 16 to write an equation: 2+6+3+◯=16. We can simplify the equation by adding 2+6+3. This gives us 11+◯=16. Since 11+⟨5⟩=16, we place a 5 in the blank hexagon.

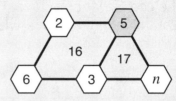

Then, we can use the region with a sum of 17 to write an equation: 3+5+*n*=17. We simplify the left side by adding 3+5. This gives us 8+*n*=17. Since 8+<u>9</u>=17, the value of *n* is **9**.

We replace *n* with 9 and check our work.

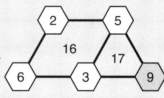

PRACTICE | Find the value of the variable in each diagram below.

77.

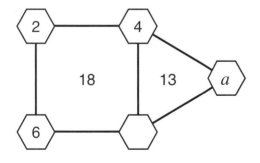

77. *a*=_____

78.

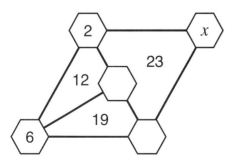

78. *x*=_____

79.

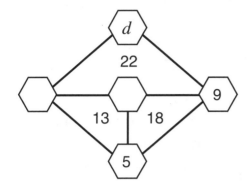

80.

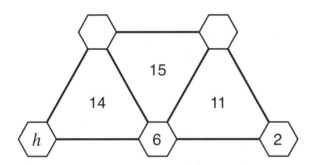

80. *h*=_____

81.

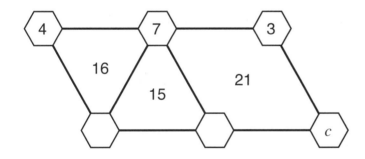

81. *c*=_____

82.

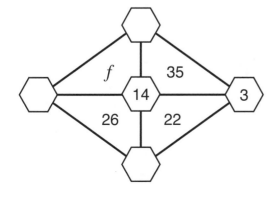

82. *f*=_____

Some of the weights on the balance scales below have been labeled with letters instead of numbers!

EXAMPLE Write an equation to represent the balance scale below. Then, solve for the unknown weight.

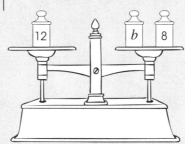

The 12-gram weight is balanced by the b-gram weight and the 8-gram weight. We can write an equation for the scale: **12=b+8**.

If we remove 8 grams from each side of the scale, the b-gram weight will balance 12−8=4 grams.

Similarly, we can solve our equation by subtracting 8 from both sides:

$$\begin{array}{r} 12=b+8 \\ \underline{-8 \quad\;\; -8} \\ 4=b \end{array}$$

So, b=**4**.

We check our answer:
12=4+8. ✓

PRACTICE Write an equation to represent each balance scale below. Then, solve for the unknown weight.

83.

83. equation: _____

$a=$_____

84.

84. equation: _____

$d=$_____

PRACTICE | Write an equation to represent each balance scale below.
Then, solve for the unknown weight.

85.

85. equation: _____

$j=$_____

86.

86. equation: _____

$n=$_____

87.

87. equation: _____

$h=$_____

88.

88. equation: _____

$w=$_____

EXAMPLE

Write an equation for the sentence below. Then solve the equation for n.

> Three more than n is 17.

"Three more than n" is $n+3$ (or $3+n$), and "is" means "equals."

So, we write the equation $n+3=17$ (or $3+n=17$).

To solve the equation, we subtract 3 from both sides:

$$\begin{array}{r} n+3=17 \\ \underline{-3 \quad -3} \\ n=14 \end{array}$$

So, $n=14$.

We replace n with 14 to check our answer:
Three more than 14 is 17. ✓

PRACTICE | Write an equation for each sentence below. Then, solve the equation for the variable.

89. The sum of six and a is thirty-six.

89. equation: _____

$a=$ _____

90. Twenty-six more than q is seventy.

90. equation: _____

$q=$ _____

91. Forty-three is the sum of w and fifteen.

91. equation: _____

$w=$ _____

92. Thirty less than t is eighty-seven.

92. equation: _____

$t=$ _____

93. Nineteen less than j is seventy-four.

93. equation: _____

$j=$ _____

PRACTICE | Write an equation for each sentence below.
Then, solve the equation for the variable.

94. Seventy-three is the sum of a number and six.
What is the number?
Use n to represent the number.

94. equation: _____

$n=$_____

95. Twelve less than a number is twenty-nine.
What is the number?
Use n to represent the number.

95. equation: _____

$n=$_____

96. Seven inches more than Grogg's height is
sixty-five inches. What is Grogg's height, in
inches? Use g to represent the number of inches
in Grogg's height.

96. equation: _____

$g=$_____

97. Six less than the number of math books is fifteen.
How many math books are there?
Use m to represent the number of math books.

97. equation: _____

$m=$_____

98. Ninety-seven is fifty-nine more than the number of
pandakeets. How many pandakeets are there?
Use p to represent the number of pandakeets.

98. equation: _____

$p=$_____

99. ★ Fifteen years ago, Devin was thirty-three years
old. How many years old is Devin today?
Use d to represent Devin's current age in years.

99. equation: _____

$d=$_____

Translating Sentences

EXAMPLE | Alex is h inches tall. If Alex grows three more inches, then he will be 32 inches tall. How many inches tall is Alex today?

If Alex grows three more inches, then he will be three inches taller than he is right now.

We write "three more than h" as $h+3$, so our equation is $h+3=32$. (We could also write $3+h=32$.)

Subtracting three from both sides, we get $h=29$.

So, Alex is **29** inches tall today.

PRACTICE | Write an equation for each word problem below. Then, solve the equation.

Many of the problems below can be solved without writing an equation. However, the problems provide good practice for writing and solving equations.

100. Ralph is r years old. In thirty-seven years, Ralph will be forty-four years old. How old is Ralph today?

100. equation: _____

$r=$_____

101. Winnie made w cookies. Alex made 36 cookies. Winnie and Alex made a total of 84 cookies. How many cookies did Winnie make?

101. equation: _____

$w=$_____

102. There are 65 adult hexatoads and h baby hexatoads in Professor Grok's office. All together, there are 122 hexatoads in his office. How many baby hexatoads are in Grok's office?

102. equation: _____

$h=$_____

103. ★ Winnie enters an elevator on floor f. She goes up 6 floors, down 4 floors, and then up 2 more floors, where she exits the elevator on floor 9. On what floor did Winnie enter the elevator?

103. equation: _____

$f=$_____

EXAMPLE | Sam is 3 years older than Tim. The sum of Sam's and Tim's ages, in years, is 17. How many years old is Tim?

We use t to represent Tim's age, in years.

Sam is 3 years older than Tim, so Sam is $t+3$ years old.

A "sum" is the result of addition, so Tim's age (t) added to Sam's age ($t+3$) is 17:

$$t+(t+3)=17.$$

The associative property of addition lets us remove parentheses in a sum, so the equation above is equal to

$$t+t+3=17.$$

Subtracting 3 from both sides, we get $t+t=14$.

Since $7+7=14$, the value of t is 7. Tim is **7** years old.

We check our answer: Tim is 7 years old. So, Sam is $7+3=10$ years old. The sum of Sam's and Tim's ages is $7+10=17$ years. ✓

PRACTICE | Solve each word problem below.

104. Together, Grogg and Alex have 32 gumballs. Grogg has twelve more gumballs than Alex. How many gumballs does Alex have?

104. _____

105. The sum of Olivia's and Ralph's heights is 64 inches. Olivia is 6 inches taller than Ralph. How many inches tall is Ralph?

105. _____

106. Together, a xylophone and case cost $100. The xylophone costs $80 more than the case. How much does the case cost?

106. _____

VARIABLES

107. Alex earned a total of $42 working after school on Monday and Tuesday. On Tuesday, he earned six dollars more than he earned on Monday. How much did Alex earn on Monday?

107.

108. Fiona scored 33 points during two Beastball games. She scored nine fewer points during the first game than she scored during the second game. How many points did she score during the second game?

108.

109. ★ Lizzie read a total of 61 pages on Wednesday, Thursday, and Friday. She read 7 pages more on Friday than she read on Wednesday. On Thursday, she read 6 pages fewer than she read on Wednesday. How many pages did Lizzie read on Wednesday?

109.

In the diagrams below, the number in each circle is the **sum** *of the numbers in the connected circles below it.*

EXAMPLE | Find the value of *j* below.

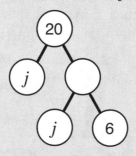

The blank circle is the sum of the two numbers connected below it, so we can label it *j*+6.

Now, we can use the top three circles to write an equation. 20 is the sum of *j* and *j*+6, so 20=*j*+(*j*+6).

Subtracting 6 from both sides of the equation gives us

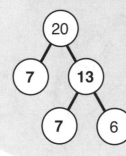

$$14 = j + j.$$

Since 7+7=14, the value of *j* is **7**.

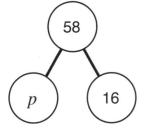

We replace *j* with 7 and check our work.

PRACTICE | Find the value of the variable in each diagram below.

110.

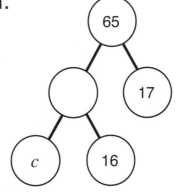

111.

110. *p*=_____

111. *c*=_____

PRACTICE | Find the value of the variable in each diagram below. When a variable is used more than once in the same problem, it represents the same number each time it is used.

112.

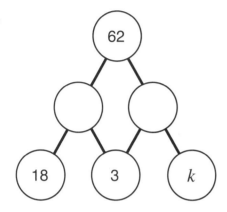

113.

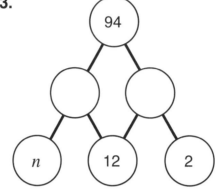

112. *k=*_____

113. *n=*_____

114.

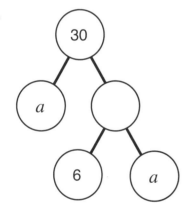

115.

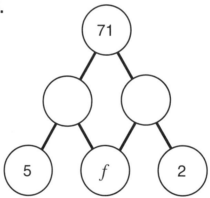

114. *a=*_____

115. *f=*_____

116.

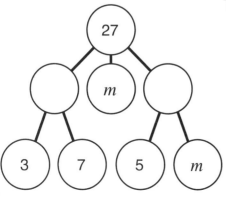

117.
★

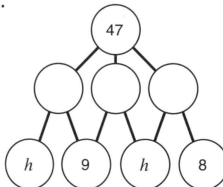

116. *m=*_____

117. *h=*_____

30

118.
★

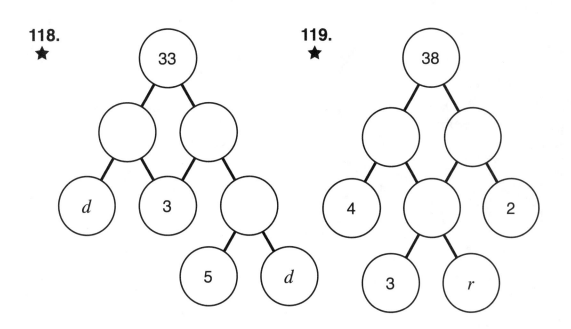

119.
★

118. *d*=_____

119. *r*=_____

120.
★

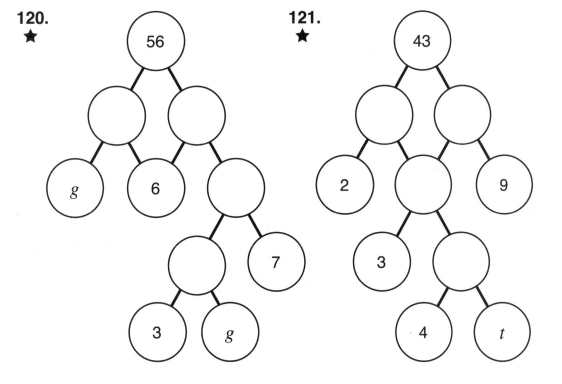

121.
★

120. *g*=_____

121. *t*=_____

EXAMPLE | Weights that are labeled with the same variable have the same weight. Solve for both unknown weights.

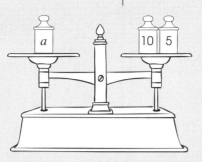 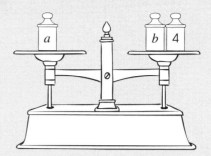

We can write an equation for each balance scale:

$a=10+5$ and $a=b+4$.

From the left scale, we get $a=10+5=15$, so $a=15$.
Since $a=15$, we can replace the a-gram weight on the right scale with a 15-gram weight, giving us $15=b+4$.

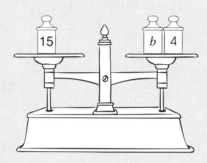

We can remove 4 grams from each side of the balance.
Similarly, we can subtract 4 from both sides of our equation.
Subtracting 4 from both sides of $15=b+4$, we get **$b=11$**.

We replace a with 15 and b with 11 to check our answers:

$15=10+5$ ✓   and $15=11+4$ ✓

PRACTICE | Weights that are labeled with the same variable have the same weight. Solve for all variables.

122.

 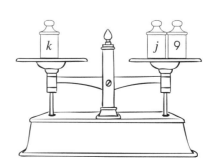

122. $j=$_____

$k=$_____

PRACTICE | Weights that are labeled with the same variable have the same weight. Solve for all variables.

123.

 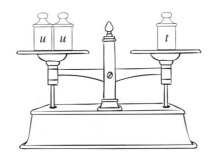

123. t=_____

u=_____

124. ★

 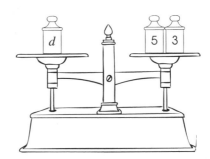

124. c=_____

d=_____

125. ★

125. g=_____

h=_____

126. ★ ★

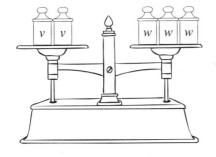 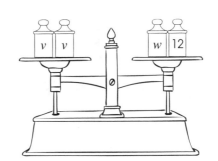

126. v=_____

w=_____

In each pair of equations below, a variable represents the same number each time it is used.

EXAMPLE | Solve for a and b.

$$a-2=10$$
$$b+6=a$$

First, we solve the equation on top: $a-2=10$.

Adding 2 to both sides, we get **$a=12$**.

Then, since $a=12$, we can replace the a in the second equation with 12:

$$b+6=12.$$

Subtracting 6 from both sides gives us **$b=6$**.

We replace a with 12 and b with 6 to check our answers:

$$12-2=10 \checkmark$$
$$6+6=12 \checkmark$$

PRACTICE | Each problem below has two equations and two variables. Solve for both variables in each problem.

127. $18= x+4$

$x= y+8$

127. $x=$_____

$y=$_____

128. $j+3= k$

$19= 12+j$

128. $j=$_____

$k=$_____

129. $c+9= 15$

$d= c+c$

129. $c=$_____

$d=$_____

130. ★ $q-6= 10$

$q+r = 7+r+r$

130. $q=$_____

$r=$_____

131. ★★ $m+m= n+4$

$4+n = 16+m$

131. $m=$_____

$n=$_____

CHAPTER 8
Division

Use this Practice book with
Guide 3C from BeastAcademy.com.

Recommended Sequence:

You may also read the entire chapter
in the Guide before beginning the
Practice chapter.

PRACTICE | Use the dot patterns to help you solve each division problem below.

1. If the 21 dots in the pattern below are divided into groups of 3, how many groups will there be?

1. _____

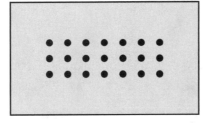

2. If the 24 dots in the pattern below are divided into groups of 4, how many groups will there be?

2. _____

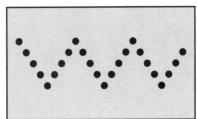

3. If the 24 dots in the pattern below are divided into groups of 8, how many groups will there be?

3. _____

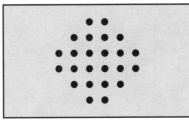

4. If the 36 dots in the pattern below are divided into groups of 9, how many groups will there be?

4. _____

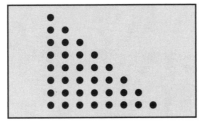

DIVISION

Grouping

If you know the total number of items, and the number of groups...

...you can divide to find the number of items in each group.

PRACTICE | Use the dot patterns to help you solve each division problem below.

5. If the 30 dots in the pattern below are divided into 6 equal groups, how many dots will be in each group?

5.

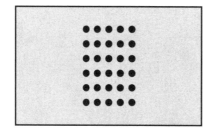

6. If the 30 dots in the pattern below are divided into 3 equal groups, how many dots will be in each group?

6.

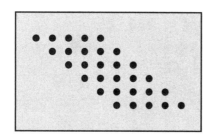

7. If the 15 dots in the pattern below are divided into 5 equal groups, how many dots will be in each group?

7.

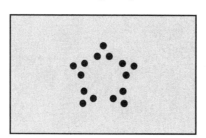

8. If the 72 dots in the pattern below are divided into 8 equal groups, how many dots will be in each group?

8.

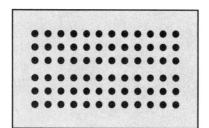

Beast Academy Practice 3C

EXAMPLES | Fill in the missing number in the equation below.

$$30 \div 5 = \boxed{}$$

We can use multiplication facts to solve division problems!

To divide $30 \div 5$, we find the number that can be multiplied by 5 to get 30.

We can use the multiplication fact $\boxed{6} \times 5 = 30$ to see that $30 \div 5 = \boxed{6}$.

PRACTICE | Fill in the missing number in each equation below.

9. $8 \times \boxed{} = 24$

10. $9 \times \boxed{} = 36$

11. $\boxed{} \times 7 = 35$

12. $5 \times \boxed{} = 10$

13. $\boxed{} \times 3 = 27$

14. $2 \times \boxed{} = 20$

15. $\boxed{} \times 6 = 42$

16. $10 \times \boxed{} = 60$

17. $35 \div 7 = \boxed{}$

18. $60 \div 10 = \boxed{}$

19. $36 \div 9 = \boxed{}$

20. $42 \div 6 = \boxed{}$

21. $10 \div 5 = \boxed{}$

22. $27 \div 3 = \boxed{}$

23. $24 \div 8 = \boxed{}$

24. $20 \div 2 = \boxed{}$

25. Connect each division problem on the right to the multiplication fact on the left that could be used to solve it.

In a division problem, the **dividend** is the number you are dividing.

The **divisor** is the number you are dividing by.

And the **quotient** is the result.

63÷9=7

PRACTICE Practice your division facts with the division wheels below.
Divide each dividend in the shaded area by the divisor in the middle.

26.

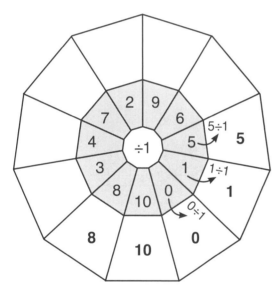

27.

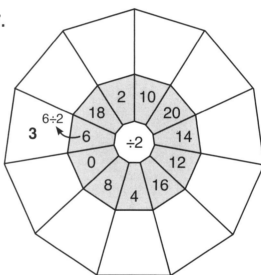

28.

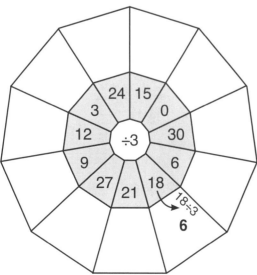

29.

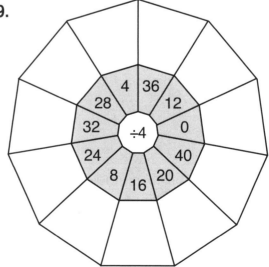

30.

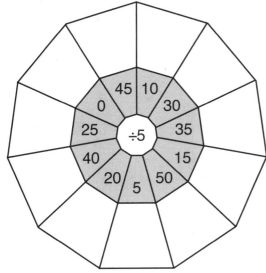

31.

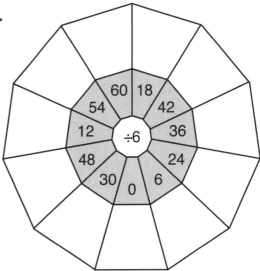

32.

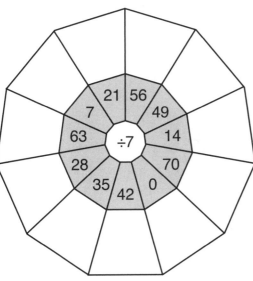

33.

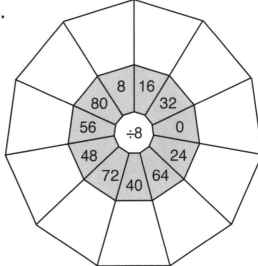

34.

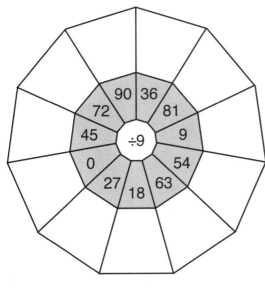

35.

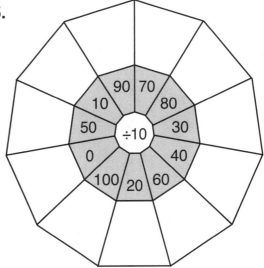

PRACTICE

Ms. Q's class plants flowers in the community garden.

36. Alex plants 100 flowers in rows, with 10 flowers in each row. How many rows of flowers does Alex plant?

36. _____

37. Winnie plants 32 flowers in 8 equal rows. How many flowers does Winnie plant in each row?

37. _____

38. Grogg plants 56 flowers in 7 equal rows. How many flowers does Grogg plant in each row?

38. _____

Lizzie's mom bakes 40 cupcakes for a birthday party at Lizzie's house.

39. If the cupcakes are divided equally among 4 little monsters, how many cupcakes will each little monster get?

39. _____

40. If the cupcakes are divided equally among 5 little monsters, how many cupcakes will each little monster get?

40. _____

41. Alex, Grogg, Lizzie, and Winnie each eat one cupcake. Then, Cammie and Ralph arrive. The remaining cupcakes are divided equally among the six little monsters. How many cupcakes does Ralph get?

41. _____

This table lists the costs of some bags of fruit at the grocery store:

Fruit	Cost of 1 Bag
Apple	$3
Banana	$4
Mango	$5
Dewberry	$8

42. How many bags of apples can Lizzie buy with $15?

42. _____

43. How many bags of bananas can Fiona buy with $28?

43. _____

44. How many bags of mangoes can Ms. Q. buy with $30?

44. _____

45. Winnie has exactly enough money to buy 5 bags of dewberries, but instead decides to spend all of her money on bananas. How many bags of bananas does Winnie buy?

45. _____

46. Alex has $24. How many **more** bags of apples can Alex buy than bags of bananas?

46. _____

47. Grogg spends $42 buying an equal number of bags of apples and bananas. How many bags of fruit does Grogg buy all together?

47. _____

You'll need to use multiplication, division, addition, and subtraction to complete these cross-number puzzles!

EXAMPLE | Complete the cross-number puzzle below.

36	÷	6	=	
÷	■	+	■	+
4	×		=	12
=	■	=	■	=
	+		=	

Across:
$36 \div 6 = \boxed{6}$.
$4 \times \boxed{3} = 12$.

Down:
$36 \div 4 = \boxed{9}$.
$6 + 3 = \boxed{9}$.

$9 + 9 = \boxed{18}$ and $6 + 12 = \boxed{18}$.

The completed cross-number puzzle looks like this:

36	÷	6	=	**6**
÷	■	+	■	+
4	×	**3**	=	12
=	■	=	■	=
9	+	**9**	=	**18**

PRACTICE | Complete each cross-number puzzle below.

48.

2	×	21	=	
×	■	÷	■	−
15	÷	3	=	
=	■	=	■	=
	+		=	

49.

56	−	50	=	
÷	■	÷	■	+
7	+	5	=	
=	■	=	■	=
	+		=	

50.

42	÷	6	=	
−	■	+	■	+
36	÷	4	=	
=	■	=	■	=
	+		=	

51.

56	−	54	=	
÷	■	÷	■	+
8	+	6	=	
=	■	=	■	=
	+		=	

PRACTICE | Complete each cross-number puzzle below.

52.

72	−	70	=	
÷	■	÷	■	+
9	+		=	16
=	■	=	■	=
	+		=	

53.

11	×	7	=	
×	■	×	■	+
6	÷		=	
=	■	=	■	=
	+	14	=	

54.

	÷	6	=	
−	■	+	■	+
32	÷		=	
=	■	=	■	=
4	+	10	=	

55.

	+	3	=	12
÷	■	−	■	÷
3	−		=	
=	■	=	■	=
	×	2	=	

56.

32	÷	4	=	
−	■	+	■	+
20	÷	2	=	
=	■	=	■	=
	+		=	

57.

	×		=	
×	■	×	■	+
4	÷	2	=	
=	■	=	■	=
32	+	10	=	

58.

64	÷		=	
−	■	+	■	+
60	÷	6	=	
=	■	=	■	=
	+	14	=	

59.

	×	6	=	36
÷	■	÷	■	÷
3	×		=	9
=	■	=	■	=
	+		=	

60.

	÷	5	=	
−	■	+	■	+
42	÷		=	6
=	■	=	■	=
3	+		=	

61.

	×	4	=	40
÷	■	÷	■	÷
5	×		=	10
=	■	=	■	=
	+		=	

62.
★

3	×	40	=	
×	■	÷	■	−
35	÷		=	
=	■	=	■	=
	+		=	

63.
★

3	×	14	=	
×	■	÷	■	−
	÷		=	
=	■	=	■	=
30	+		=	

Try dividing some larger numbers.

EXAMPLE | What is 320÷8?

To divide 320÷8, we find the number that can be multiplied by 8 to get 320.

Since 4×8=32,

$\boxed{40}$×8=320.

So, 320÷8=**40**.

PRACTICE | Find each quotient.

64. 500÷5=_____

65. 630÷7=_____

66. 400÷8=_____

67. 900÷3=_____

68. 2,400÷4=_____

69. 1,800÷9=_____

70. 42,000÷6=_____

71. 20,000÷5=_____

72.
★ 18,000÷20=_____

73. 14,000÷700=_____
★

Long Division

Long division is used to compute a quotient and remainder!

Review long division on page 60 of the Guide.

EXAMPLE | Find the quotient and remainder of 73÷6.

First, we write the division problem as shown:

6) 73

The **dividend** goes here... ...and the **divisor** goes here. The **quotient** goes here.

6×10=60, so 6 can go into 73 at least ten times.

```
  10
6) 73
  -60
   13
```

We subtract 6×10=60 from 73, and have 13 left over. The remainder cannot be greater than the divisor. Since 13 is greater than 6, we keep dividing.

Six goes into 13 two times, so we add two to the quotient.

```
  10+2
6) 73
  -60
   13
  -12
    1
```

We subtract 2×6=12 from 13 and have 1 left over. Since 1 is less than 6, the remainder is 1.

The quotient of 73÷6 is 10+2=**12**, and the remainder is **1**.

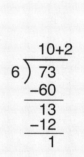

PRACTICE | Find the quotient and remainder of each problem below.

74. 5) 67

quotient=_____

remainder=_____

75. 3) 43

quotient=_____

remainder=_____

76. 7) 64

quotient=_____

remainder=_____

77. 4) 76

quotient=_____

remainder=_____

78. $8 \overline{)\ 93}$ quotient=_____

remainder=_____

79. $7 \overline{)\ 98}$ quotient=_____

remainder=_____

80. $11 \overline{)\ 13}$ quotient=_____

remainder=_____

81. $91 \overline{)\ 95}$ quotient=_____

remainder=_____

82. $9 \overline{)\ 5}$ quotient=_____

remainder=_____

83. $6 \overline{)\ 125}$ quotient=_____

remainder=_____

84. ★ $14 \overline{)\ 86}$ quotient=_____

remainder=_____

85. ★ $8 \overline{)\ 2507}$ quotient=_____

remainder=_____

86. ★ When Winnie divides n by 8, the quotient is 15 and the remainder is 6. What is n?

86. _____

87. ★ When Lizzie divides 77 by m, the quotient is 9 and the remainder is 5. What is m?

87. _____

For each of the following Remainder Jump mazes, begin in the hexagon marked "Start." Divide the number in the hexagon by the given divisor. The remainder is the exact number of spaces you must move to reach a new hexagon.

Divide the number in the new hexagon by the given divisor and move again.

Continue in this way until you land on the hexagon marked "Finish." You may not cross or land on the same hexagon twice.

Trace your final path. Circle each number you land on.

EXAMPLE | *Divisor:* 4

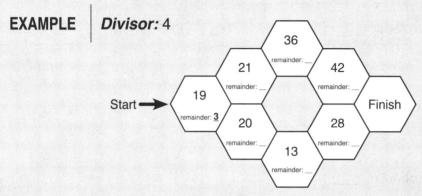

We begin by dividing 19÷4. Since 19÷4 has remainder **3**, we must move 3 spaces. There are only four possible moves:

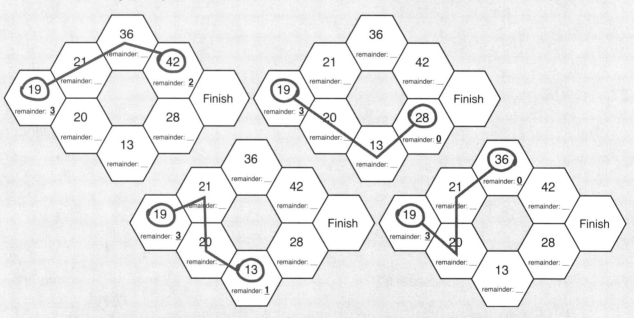

Only landing in the hexagon marked 42 allows us to continue to the hexagon marked "Finish." 42÷4 has remainder **2**. We move 2 spaces as shown to land on the hexagon marked "Finish."

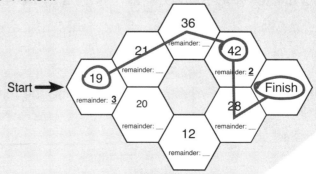

PRACTICE | There is only one correct path to each "Finish" hexagon.

88. *Divisor:* 10

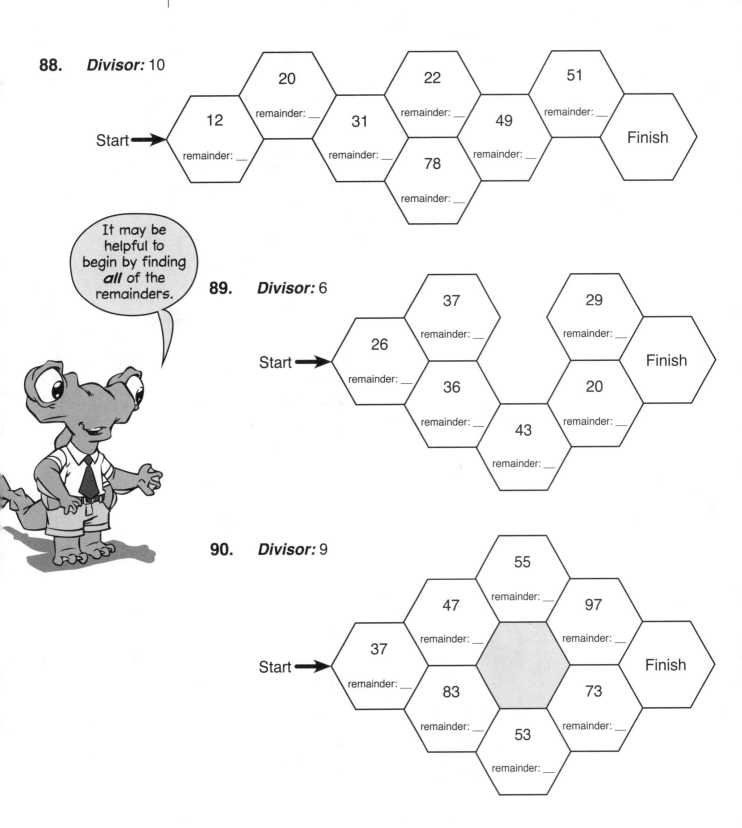

It may be helpful to begin by finding *all* of the remainders.

89. *Divisor:* 6

90. *Divisor:* 9

PRACTICE | There is only one correct path to each "Finish" hexagon.

91. ★ **Divisor:** 5

92. ★ **Divisor:** 8

93. ★ **Divisor:** 7

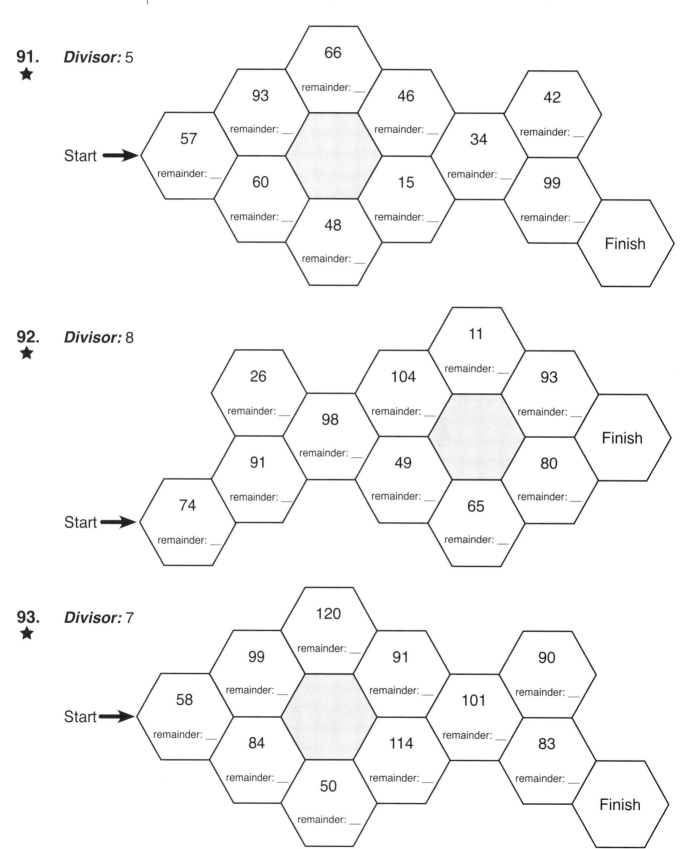

94. ⭐ *Divisor:* 5

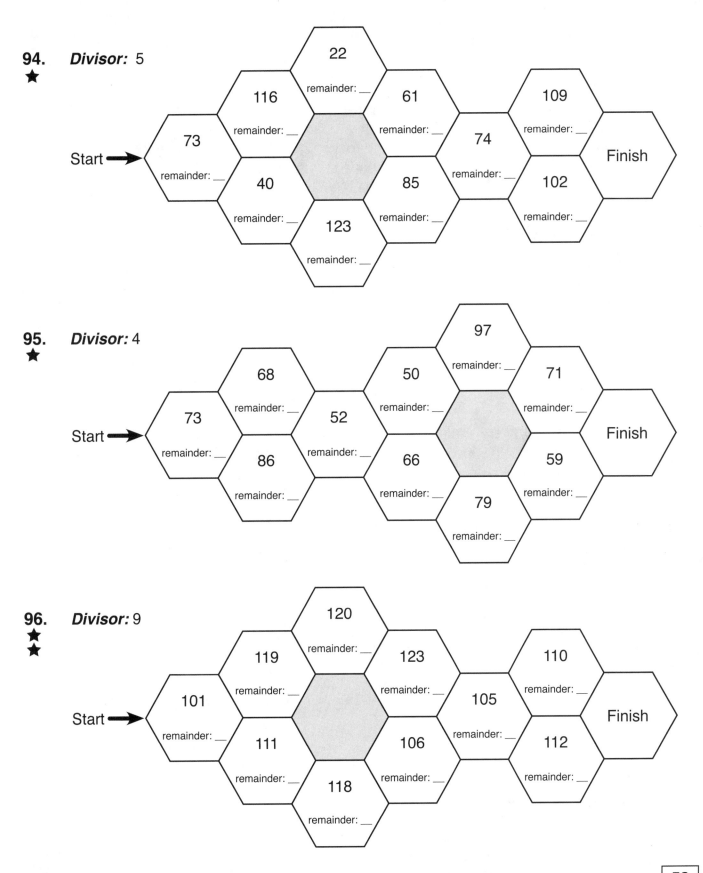

95. ⭐ *Divisor:* 4

96. ⭐⭐ *Divisor:* 9

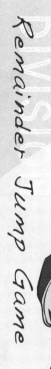

Who'll be first to reach the finish?

Materials:
Game board, cards numbered 2 through 9 (these can be borrowed from a standard deck of cards), and a small game piece for each player. Sample game boards are on the next two pages. You can print more game boards at BeastAcademy.com, or make your own.

Players:
2 or more.

Object:
Be the first player to reach the "Finish" hexagon.

Game Play:
The numbered cards are shuffled and placed face-down in a pile. All players begin with their game pieces in the hexagon marked "Start." Players take turns, starting with the youngest player and continuing to the left.

On a player's turn, he or she selects a card from the top of the deck. The player divides the number in the hexagon of his or her game piece by the number on the card.

The remainder is the number of spaces a player may move his or her game piece to reach a new hexagon. Players may move in any direction, but may not visit the same hexagon twice in the same turn.

Players continue drawing cards, dividing, and moving to new hexagons. The first player to end his or her turn on the "Finish" hexagon wins the game!

Find more game boards and strategy hints at BeastAcademy.com!

DIVISION

Start

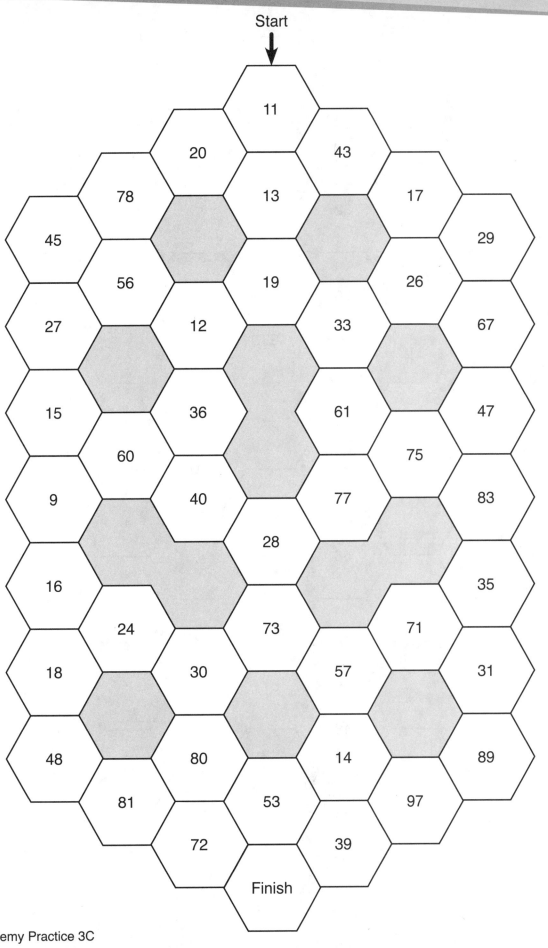

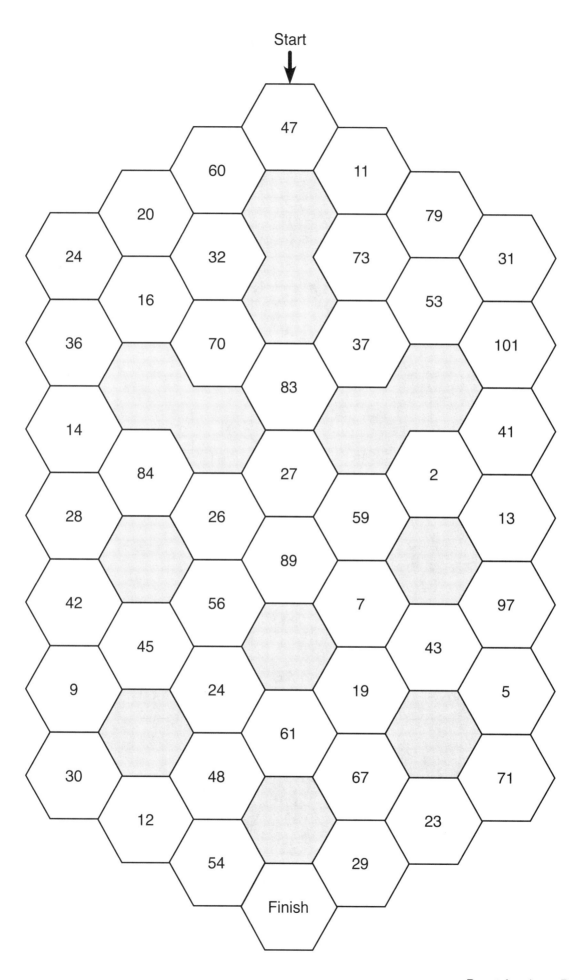

EXAMPLE | What is the side length of a regular pentagon with a perimeter of 115?

A regular pentagon has 5 sides of equal length. So, to find the length of each side, we divide 115 by 5.

Division can also help us answer questions about perimeter and area.

$$\begin{array}{r} 20+3 \\ 5{\overline{\smash{)}\,115}} \\ -100 \\ \hline 15 \\ -15 \\ \hline 0 \end{array}$$

Since the remainder is 0, we can write

$$115 \div 5 = 20 + 3 = 23.$$

So, the side length of a regular pentagon with perimeter 115 is **23**.

PRACTICE | The areas given below are in square units.

97. What is the side length of a regular hexagon with a perimeter of 156?

97. _____

98. What is the height of a rectangle with an area of 119 and width 7?

98. _____

99. What is the side length of a regular octagon with a perimeter of 184?

99. _____

100. Ralph draws a regular polygon with perimeter 108 and side length 9. How many sides does Ralph's polygon have?

100. _____

The areas given below are in square units.

EXAMPLE

A rectangle with height 5 and width 18 has an area of 90. What is the width of a rectangle with the same area and height 18?

Try these!

To find the area of a rectangle, we can multiply its height by its width. So, to find the width of a rectangle, we can divide its area by its height: 90÷18.

To divide 90÷18, we find the number that can be multiplied by 18 to get 90.

Since the first rectangle has height 5, width 18, and area 90, we know 5×18=90. So, 5 can be multiplied by 18 to get 90. Therefore, 90÷18=5.

So, a rectangle with an area of 90 and height 18 has width **5**.

PRACTICE

101. The rectangle below has area 105 and height 7. What is the width of the rectangle?

101. _____

7

102. A rectangle has area 105 and height 15. What is the width of the rectangle?

102. _____

Use the following for Problems 103 and 104:
Lizzie and Grogg each have 117 markers.

103. Lizzie's markers are organized into packs of 9. How many packs of markers does Lizzie have?

103. _____

104. Grogg's markers are stuffed into packs of 13. How many packs of markers does Grogg have?

104. _____

105. If Captain Kraken divides 322 coins from a treasure chest into piles of 14, he can make 23 piles. How many piles of coins can Captain Kraken make if he divides the 322 coins into piles of 23?

105. _____

106. There are 7 days in a week, and 24 hours in each day. So, there are 7×24=168 hours in one week. How many weeks are there in 168 days?

106. _____

107. If $n \div 4 = 5$, what is $n \div 5$?

107. _____

108. ★ If $n \div 151 = 239$, what is $n \div 239$?

108. _____

109. ★ If $64 \div m = n$, what is $64 \div n$?

109. _____

110. ★ If $a \div b = c$, what is $a \div c$? (Assume that a, b, and c are not zero.)

110. _____

EXAMPLE | What is the remainder when 30+31+32 is divided by 7?

We first add 30+31+32, then divide by 7 to find the remainder.
30+31+32=93.

$$\begin{array}{r} 10+3 \\ 7{\overline{\smash{\big)}\,93}} \\ -70 \\ \hline 23 \\ -21 \\ \hline 2 \end{array}$$

93÷7 has remainder **2**.

— *or* —

First, we find the remainder when each number is divided by 7.

30÷7 has remainder 2.

31÷7 has remainder 3.

32÷7 has remainder 4.

Then, we add the remainders.
(30+31+32)÷7 has the same remainder as (2+3+4)÷7.

2+3+4=9, and 9÷7 has remainder 2.
So, (30+31+32)÷7 has remainder **2**.

PRACTICE | Find the *remainder* for each division problem below.

111. Alex has 74 green buttons and 75 blue buttons. When he arranges the buttons into rows of 7, how many buttons will be left over?

111. _____

112. (36+37)÷5

113. (39+40)÷6

112. remainder=_____

113. remainder=_____

114. (94+95+96)÷9

115. (46+47+48)÷4

114. remainder=_____

115. remainder=_____

116. (13+14+15)÷12

117. (187+188+189)÷75
★

116. remainder=_____

117. remainder=_____

EXAMPLE | What is the remainder when 10×18 is divided by 7?

We first multiply 10×18, then divide by 7 to find the remainder.
10×18=180.

$$\begin{array}{r} 20+5 \\ 7\overline{)180} \\ -140 \\ \hline 40 \\ -35 \\ \hline 5 \end{array}$$

180÷7 has remainder **5**.

— *or* —

First, find the remainder when each number is divided by 7.

10÷7 has remainder 3.

18÷7 has remainder 4.

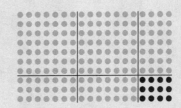

Then, we multiply the remainders.
(10×18)÷7 has the same remainder as (3×4)÷7.

3×4=12, and 12÷7 has remainder 5.
So, (10×18)÷7 has remainder **5**.

PRACTICE | Find the *remainder* for each division problem below.

118. The Beast Bakery orders 12 boxes of 8 lemons to make pies. The bakers use 7 lemons in each pie. After they make as many pies as possible, how many lemons are left over?

118. _____

119. (10×11)÷8

120. (57×58)÷6

119. remainder=_____

120. remainder=_____

121. (54×55)÷53

122. (6×7×8)÷5

121. remainder=_____

122. remainder=_____

123. (11×11×11)÷9

124. (103×104×105)÷100
★

123. remainder=_____

124. remainder=_____

EXAMPLE | Place the given numbers into the circle diagram so that the sum of the two numbers in each circle has remainder 0 when divided by 4.

Numbers to place: 7, 13, 21

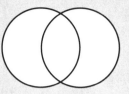

We can place the numbers as shown below:

The left circle contains 7 and 13.
7+13=20, which has remainder 0 when divided by 4.
The right circle contains 7 and 21.
7+21=28, which has remainder 0 when divided by 4.

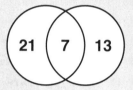

You may have switched the 13 and the 21, as shown.

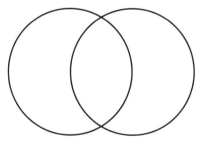

PRACTICE | Place the given numbers into the circle diagram so that the sum of the two numbers in each circle has remainder 0 when divided by 5.

125. Numbers to place: 12, 17, 23

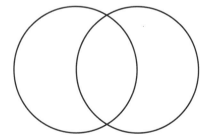

126. Numbers to place: 11, 24, 39

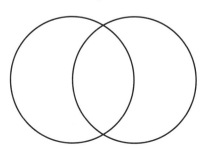

127. Numbers to place: 13, 23, 27

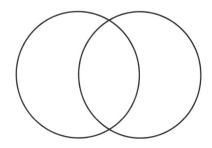

128. Numbers to place: 36, 31, 49

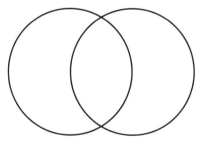

PRACTICE | Place the given numbers into the circle diagram so that the sum of the two numbers in each circle has remainder 0 when divided by 8.

129. ***Numbers to place:*** 15, 23, 65

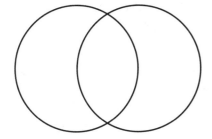

130. ***Numbers to place:*** 11, 21, 29

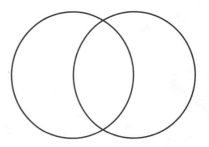

131. ***Numbers to place:*** 38, 46, 58

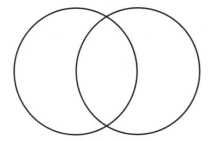

132. ***Numbers to place:*** 65, 73, 79

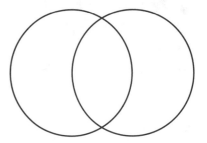

PRACTICE | Place the given numbers into the circle diagram so that the sum of the two numbers in each circle has remainder ***3*** when divided by 7.

133. ***Numbers to place:*** 64, 71, 79

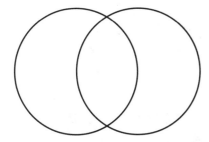

134. ***Numbers to place:*** 32, 48, 62

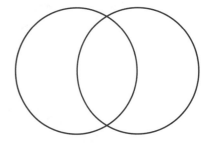

135. ***Numbers to place:*** 85, 93, 107

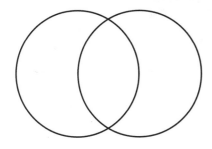

136. ***Numbers to place:*** 81, 95, 111

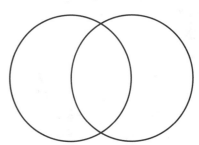

EXAMPLE | Complete each diagram with the given numbers so that no two connected circles have a sum that has remainder 0 when divided by 5.

Missing Numbers: 9, 10

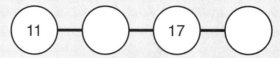

9+11=20, which has remainder 0 when divided by 5, so 9 cannot be placed in the empty circle next to 11. This leaves only one possible circle for the 9:

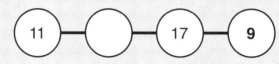

The 10 can be placed in the remaining circle.

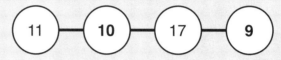

Each of 11+10, 10+17, and 17+9 has a remainder that is not 0 when divided by 5.

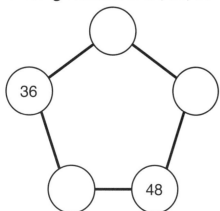

PRACTICE | Complete each diagram with the given numbers so that no two connected circles have a sum that has remainder 0 when divided by 10.

137. ***Missing Numbers:*** 15, 16, 17

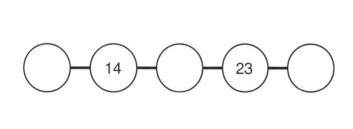

138. ***Missing Numbers:*** 21, 22, 23

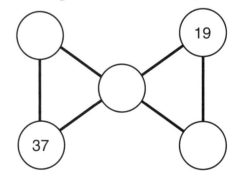

139. ***Missing Numbers:*** 52, 53, 54

140. ***Missing Numbers:*** 19, 20, 21

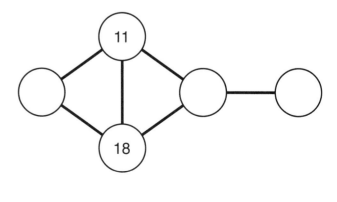

Beast Academy Practice 3C

PRACTICE | Complete each diagram with the given numbers so that no two connected circles have a sum that has remainder 0 when divided by 5.

141. *Missing Numbers:* 26, 27, 28

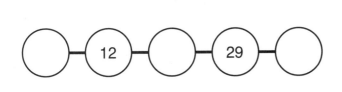

142. *Missing Numbers:* 32, 33, 34

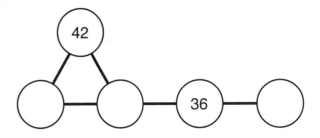

143. *Missing Numbers:* 15, 16, 17

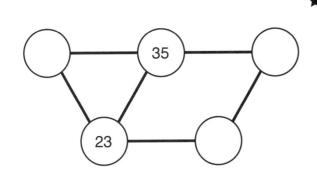

144. *Missing Numbers:* 41, 42, 43
★

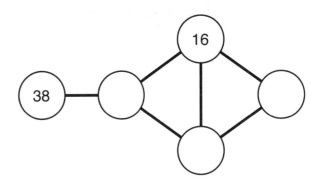

PRACTICE | Complete each diagram with the given numbers so that no two connected circles have a sum that has remainder 0 when divided by 6.

145. *Missing Numbers:* 32, 33, 34
★

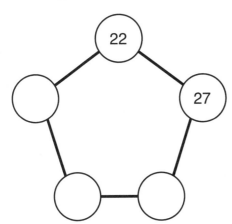

146. *Missing Numbers:* 71, 72, 73
★

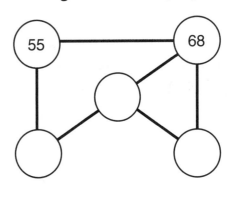

PRACTICE

147. When 234 little monsters are divided equally into 9 classrooms, how many little monsters are in each classroom?

147. _____

148. Lizzie has 7 pages of stickers, with 9 stickers on each page. If Lizzie divides her stickers equally onto 9 pages, how many stickers will there be on each page?

148. _____

149. A regular nonagon (9 sides) has sides of length 13. What is the side length of an equilateral triangle that has the same perimeter as the nonagon?

149. _____

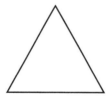

150. Grogg has a box that contains only red and purple crayons. For every red crayon in the box, there are two purple crayons. If the box holds a total of 78 crayons, how many of the crayons are purple?

150. _____

151. Captain Kraken has 7 bags of rubies, with 36 rubies in each bag. If Captain Kraken divides his rubies equally among 4 treasure chests, how many rubies will there be in each chest?

151. _____

152. Ms. Q. has 6 packs of pencils. Each pack contains 20 pencils. If Ms. Q. divides the pencils equally among the 17 students in her class, how many pencils will she have left over?

152. _____

153. A rectangle of width 16 and height 9 is divided into nine equal rectangles as shown below. What is the area of one of the small rectangles?

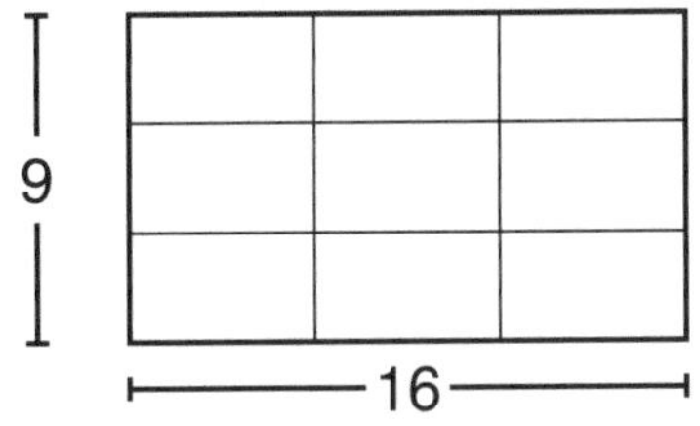

153. _____

154. Grogg can divide his gumballs into 6 piles, with 17 gumballs in each pile. If Grogg divides his gum into just 3 piles, how many gumballs will there be in each pile?

154. _____

155. When 78 blocks are stacked in piles of 7, they form 11 complete stacks with 1 block left over. If 78 blocks are stacked in piles of 14, how many blocks will be left over?

155. _____

156. Ms. Q. divides the students in her class into 7 teams, with 5 students on each team. She then divides 210 toothpicks equally among all of the students in her class. How many toothpicks does each student get?

156. _____

157. Grogg brings 50 blue, 58 purple, 66 green, and 74 pink popsicle sticks to art class. It takes 7 popsicle sticks to make a picture frame. How many popsicle sticks will be left over after Grogg makes as many picture frames as possible?

157. _____

158. ★ Alex is decorating cookies. Each cookie gets exactly 3 chocolate chips and 4 cinnamon candies. Alex has 100 chocolate chips and 125 cinnamon candies. How many cookies can Alex decorate?

158. _____

159. One school bus can hold 40 little monsters. How many buses are
★ needed to take 223 little monsters on a field trip?

159. _____

160. When 41 is divided by 7, the quotient is a with remainder b.
★ What number can be divided by 7 to get quotient b with remainder a?

160. _____

161. There are 366 days in a leap year. Kara was born Sunday,
★ January 1st during a leap year. On which day of the week will
Kara celebrate her first birthday?

161. _____

162. Grogg writes his name many times all over a sheet of paper. When
★ he is finished, he counts 162 G's on the sheet. How many O's are on
the sheet of paper?

162. _____

CHAPTER 9
Measurement

Use this Practice book with
Guide 3C from BeastAcademy.com.

Recommended Sequence:

For Chapter 9, we recommend that
you read the entire chapter in the
Guide before beginning the Practice
chapter.

Length, weight, volume, temperature, price, and time are all measures.

Measures help us describe people, places, objects, and events with numbers.

Length describes how long something is. **Distance** describes how far things are apart. Length and distance can be measured using a ruler or a tape measure.

Weight describes how heavy something is. Weight can be measured using a scale.

Volume describes how much space something takes up. Volume can be measured using a measuring cup or a graduated cylinder.

Temperature describes how hot something is. Temperature can be measured using a thermometer.

Price tells us how much something costs.

Time describes how long something takes to happen. Time can be measured using a clock or a stopwatch.

PRACTICE | Connect each item on the left with **all** of the measures on the right that are useful for describing the item.

1. A kite string

2. A refrigerator

3. A footrace

4. A swimming pool

5. A movie

Length

Weight

Volume

Temperature

Price

Time

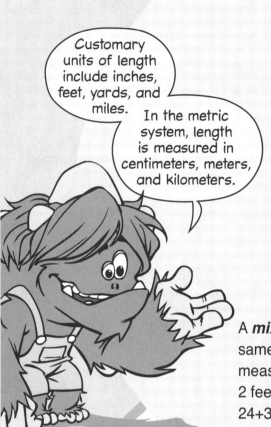

Customary units of length include inches, feet, yards, and miles.

In the metric system, length is measured in centimeters, meters, and kilometers.

Units of Length and Distance

Abbreviations for each unit are in parentheses.

Customary Unit	Conversion
inch (in)	
foot (ft)	1 ft = 12 in
yard (yd)	1 yd = 3 ft
mile (mi)	1 mi = 1,760 yd = 5,280 ft

Metric Unit	Conversion
centimeter (cm)	
meter (m)	1 m = 100 cm
kilometer (km)	1 km = 1,000 m

A **mixed measure** includes two different units from the same system. For example, 2 feet 3 inches is a mixed measure. Since 2 feet equals 2×12=24 inches, 2 feet 3 inches means the same thing as 24+3=27 inches.

PRACTICE | Use the information above to help you solve each problem below.

6. Barry is 7 feet 3 inches tall. What is Barry's height in inches?

6. _____

7. How many centimeters are there in three and a half meters?

7. _____

8. Lizzie cuts a five-foot rope into three equal pieces. How many inches long is each piece of rope?

8. _____

9. The perimeter of a square is one meter. What is the length in centimeters of one side of the square?

9. _____

PRACTICE | Use the information on the previous page to help you solve each problem below.

10. Three laps around a fitness track equals a length of one mile. What is the length in feet of one lap around the fitness track?

10. _____

11. ★ If seven blocks can be stacked to a height of 40 centimeters, how many blocks will it take to make a stack that is 2 meters tall?

11. _____

12. ★ The width of a rectangle is double its height. Its perimeter is 54 inches. What is its height?

12. _____

13. ★ Lizzie is stacking cups. Each cup is 6 inches tall. Two stacked cups reach a height of 8 inches. How many cups must Lizzie stack to make a tower that is 3 feet tall?

13. _____

6 in   8 in

14. ★ Order the following distances from longest to shortest:

1 mi   30 yd   100 ft   1,000 in

14. _____

_____

_____

_____

We use a ruler to measure length in inches and centimeters.

Small objects can be measured with a **ruler**. Most rulers have measurements on both sides. One side is used for measuring length in **inches**, and the other is used for measuring length in **centimeters**.

To measure an object, place the end of the ruler marked with a zero at one end of the object. The length of the object is indicated by the number reached by the other end of the object.

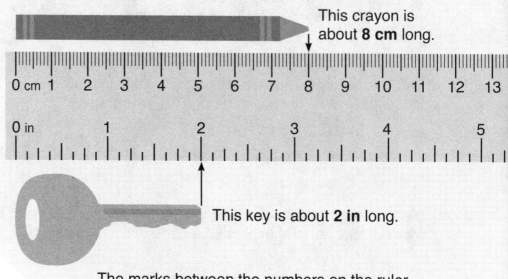

This crayon is about **8 cm** long.

This key is about **2 in** long.

The marks between the numbers on the ruler represent fractions of inches and centimeters. We will discuss these marks in Beast Academy 3D.

PRACTICE | Use the four points below to answer the questions that follow.

• A   • B   • C   • D

15. What is the distance in inches between points B and C?   15. _____

16. Which two points are 4 inches apart?   16. _____ and _____

17. Which two points are 5 inches apart?   17. _____ and _____

18. Which two points are 6 inches apart?   18. _____ and _____

PRACTICE Bobby the bacterium has friends **E**d, **F**ran, **G**il, **H**ank, **J**o, and **K**iki. Each lives in a tiny house marked by the first letter of his or her name. Use the six labeled houses below to answer the questions that follow.

E   F   G   H   J   K

19. What is the distance in centimeters from Fran's house to Jo's house?

19. _____

20. Bobby begins at Ed's house and visits Gil, then Fran, then Hank, then Kiki, then Jo. How many centimeters does Bobby travel?

20. _____

21. For each distance below, find two houses that are separated by that distance. The first answer is given.

1 cm: __J__ and __K__   2 cm: _____ and _____   3 cm: _____ and _____

4 cm: _____ and _____   5 cm: _____ and _____   6 cm: _____ and _____

7 cm: _____ and _____   8 cm: _____ and _____   9 cm: _____ and _____

22. ★ In Problem 21, you found two houses that are separated by each distance from 1 cm to 9 cm. What is the smallest whole number of centimeters for which no two houses are separated by that distance (not including 0 cm)?

22. _____

23. ★★ If Bobby the bacterium visits Ed, then Fran, then Gil, then Hank, then Jo, then Kiki, he will need to travel 17 cm. List the friends in an order that will require Bobby to travel 62 cm to visit all six friends in that order.

23. _____ _____ _____ _____ _____ _____

EXAMPLE

What is the perimeter, in **centimeters**, of the rectangle below?

We can use a ruler to measure the perimeter of a polygon.

We can use a ruler to measure the height and width of the rectangle. The rectangle is 2 cm tall and 3 cm wide, so its perimeter is 2+3+2+3=**10 centimeters**.

0 cm   1   2   3

PRACTICE | Find the perimeter in **centimeters** of each polygon below.

24. Regular Pentagon

25. Rectangle

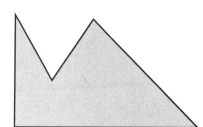

24. _____

25. _____

26. Right Triangle

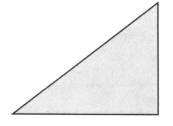

27. Pentagon

26. _____

27. _____

Use points A, B, C, and D below to answer the questions that follow.

●
A

●
B

●
D

●
C

28. What is the perimeter in centimeters of quadrilateral ABCD?

28. _____

29. What is the perimeter in centimeters of triangle ACD?

29. _____

30. What is the greatest possible perimeter in centimeters of a triangle formed by connecting three of the points above?

30. _____

Use points E, F, G, and H below to answer the questions that follow.

●
E

●
H

●
G

●
F

31. What is the perimeter in centimeters of quadrilateral EFGH?

31. _____

32. Which three points can be connected to make a triangle with a perimeter of 13 centimeters?

32. _____ _____ _____

33. What is the greatest possible perimeter in centimeters of a triangle formed by connecting three of the points above?

33. _____

A **pendulum** is a suspended weight that swings freely.
One hundred years ago, the best clocks used pendulums to keep time.
In this project, we will explore several properties of pendulums that make them excellent timekeepers.

Complete this project with the help of an adult.

Step 1: Make a pendulum.
You will need: tape, thin string (dental floss works great), a small weight (like a C battery), a stopwatch, and a tall person.
1. Cut a string between 100 and 200 centimeters in length.
2. Tape one end of the string to the small weight.
3. Tape the other end of the string to the top of an open doorway. This is where the tall person comes in handy.

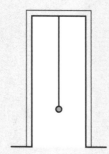

Step 2: Explore.
Your goal is to time the swings of pendulums of different lengths.
1. Measure the length of your pendulum in centimeters. The length of your pendulum is the distance from the top of the string to the middle of the weight at the end.
2. Use a stopwatch to measure your pendulum's period. The **period** is the time your pendulum takes to make one full (out-and-back) swing.
3. Change the length of your pendulum. Repeat steps 1 and 2 for pendulums of several different lengths. Create a table like the one below to record your results. We filled in one entry for you.

Length (cm)	Period (seconds)
140	2.37

Step 3: Answer the following questions based on your experiments.

1. Does pulling the pendulum farther back change its period?
2. Does changing the length of the pendulum change its period?
3. What is the length in centimeters of a pendulum that has a period of 2 seconds?

Answers are in the solutions section after Problem 33.

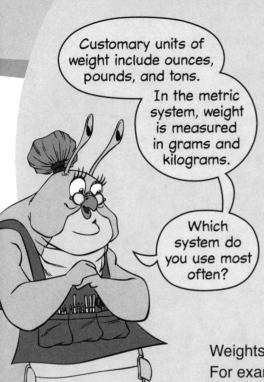

Customary units of weight include ounces, pounds, and tons.

In the metric system, weight is measured in grams and kilograms.

Which system do you use most often?

Units of Weight

Customary Unit	Conversion
ounce (oz)	
pound (lb)	1 lb = 16 oz
ton	1 ton = 2,000 lb

A large chicken egg weighs about 2 ounces.

Metric Unit	Conversion
gram (g)	
kilogram (kg)	1 kg = 1,000 g

A nickel weighs exactly 5 grams.

Weights are often listed as mixed measures.
For example, 5 pounds 7 ounces equals 87 ounces.
Since 5 pounds equals 5×16=80 ounces,
5 pounds 7 ounces equals 80+7=87 ounces.

PRACTICE | Use the information above to help you solve each problem below.

34. How many grams are there in four kilograms?

34. _____

35. When Alex was born, he weighed 3 pounds 9 ounces. What was Alex's birth weight in ounces?

35. _____

36. At the Academy Steak House, you can order a 50-ounce steak. It comes with a 14-ounce baked potato. How many **pounds** does the whole meal weigh?

36. _____

37. ★ Lizzie has five textbooks. Each book weighs 1 lb 5 oz. What is the total weight of Lizzie's books in pounds and ounces? *When a weight is given in pounds and ounces, the number of ounces must always be less than 16.*

37. _____lb _____oz

> A scale is used to measure weight.

A **spring scale** uses a spring that is stretched or compressed. On the spring scale to the right, the weight is displayed in pounds and ounces. The numbered tick marks on the scale indicate pounds. The tick marks between the numbered tick marks indicate ounces.

The arrow on this spring scale points at the 11th tick mark past 3, so the sack on this spring scale weighs 3 lb 11 oz.

A **balance scale** compares weights. The object you are weighing is placed on one side of the scale. Objects of known weight are placed on the other side until the two sides are level.

When the two sides are level, we say that the scale is balanced, meaning the weight on each side is the same. The elefinch on this balance scale weighs 50+40=90 grams.

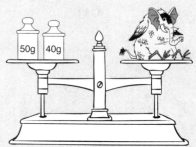

PRACTICE | Write the weight displayed on each scale below in the units listed.

38.

39.

38. _____ lb _____ oz

39. _____ g

40.

41.

40. _____ g

41. _____ lb _____ oz

PRACTICE

42. What is the combined weight in pounds and ounces of the two sacks of coins on the scales below?

42. _____lb _____oz

43. Use your answers from the previous page for the weight of the octapug and the pandakeet. How can all five items below be placed on the scale so that the scale is balanced?

23g   15g   10g

Octapug   Pandakeet

44. What is the weight of one scale?

44. _____lb _____oz

Units of Volume and Capacity

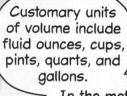

Customary units of volume include fluid ounces, cups, pints, quarts, and gallons.

In the metric system, volume is measured in milliliters and liters.

Customary Unit	Conversion
fluid ounce (fl oz)	
cup (c)	1 c = 8 fl oz
pint (pt)	1 pt = 2 c
quart (qt)	1 qt = 2 pt
gallon (gal)	1 gal = 4 qt

Metric Unit	Conversion
milliliter (mL)	
liter (L)	1 L = 1,000 mL

Volume is how much space something takes up.

Capacity is how much volume a container can hold.

—1 fl oz—

1 mL

1 cm

A milliliter is a very small amount, equal to the volume of a cube that has 1 cm edges.

PRACTICE

45. How many fluid ounces are in a pint? A quart? A gallon?

1 pt = _____ fl oz

1 qt = _____ fl oz

1 gal = _____ fl oz

46. How many milliliters are there in two and a half liters?

46. _____

47. How many fluid ounces are in a half-gallon plus a half-pint?

47. _____

48. A glass of milk holds 10 fluid ounces. How many *pints* of milk will it take to fill 8 of these glasses?

48. _____

We can measure volume with a measuring cup or a graduated cylinder.

EXAMPLE | What is the volume of water in each container below?

The graduated cylinder on the left is marked with milliliters. It contains **29 mL** of water.

This measuring cup is marked in cups and fluid ounces. It contains **11 fl oz** of water. This can also be written as a mixed measure. Since 1 cup equals 8 ounces, 11 fluid ounces is 3 fluid ounces more than 1 cup. So, 11 oz=**1 cup 3 oz**.

PRACTICE | Write the volume of the water in each measuring cup or graduated cylinder below. Use the units listed.

49. _____mL

50. ___ cup ___ fl oz

51. If the liquid in the three measuring cups below is used to fill as many 5-fluid-ounce jars as possible, how many fluid ounces of water will be left over?

51. _____

Archimedes was a Greek mathematician who lived more than 2,000 years ago.

There is a famous story of one of Archimedes's earliest discoveries. Archimedes needed to find the volume of a crown. One day, while taking a bath, Archimedes noticed that as he sunk his body into the water, the water level in the tub would rise. Archimedes realized that the change in the level of the water in the tub could be used to determine the volume of an object.

Upon making this discovery, Archimedes leaped from the tub without remembering to put his clothes on and ran through the streets shouting "Eureka!" which means "I've found it!"

While no one knows for sure if the *story* is real, Archimedes's discovery gives us a great way to find the volume of an object.

Goal: Find the volume of several objects.
You will need a measuring cup, a pitcher, a large bowl, and several objects that fit in the pitcher (examples are given below).

Step 1: Setup.
Fill a pitcher until it overflows.
Place the full pitcher inside a large empty bowl.

Step 2: Sink something.
Place a waterproof object into the pitcher so that it is completely under water. The object will cause some of the water in the pitcher to spill out into the bowl.

Step 3: Measure the volume of the spilled water.
Pour the water that spilled from the pitcher into a measuring cup or graduated cylinder. The volume of the spilled water is equal to the volume of the object!

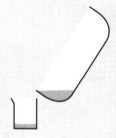

Step 4: Try some other objects!
A coffee mug, a flip-flop, a softball, an action figure, your hand, your dad's hand, etc. Waterproof items only, please.

Step 5: Answer the following questions based on your experiments.
1. How can you measure the volume of an object that floats?
2. How could you measure the volume of an object without spilling any water?
3. How could you measure the volume of your entire body without spilling any water?

Answers are in the solutions section after Problem 51.

Temperature is measured in **degrees**.

The customary unit of temperature is the **degree Fahrenheit (°F)**.

The metric unit of temperature is the **degree Celsius (°C)**.

We use a **thermometer** to measure temperature.

A comfortable room temperature is 23°C, or about 73°F.

At 15°C, which is 59°F, you would wear a jacket.

Water freezes at 0°C, which is 32°F.

Water boils at 100°C, which is 212°F.

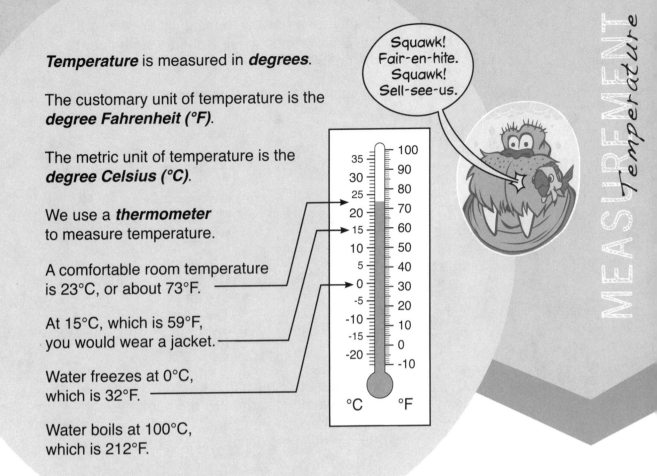

Squawk! Fair-en-hite. Squawk! Sell-see-us.

PRACTICE | Draw a line to connect each activity on the left with the arrow indicating an appropriate outdoor temperature on the right.

52. Snowball fight!

53. Swimming at the beach.

54. Picnic in the park.

55. Evening campfire.

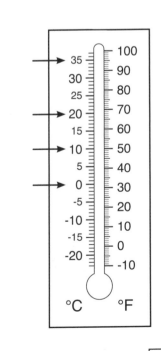

Money on Beast Island has the same units as money in the United States. Units of money are **dollars ($)** and **cents (¢)**. There are 100 cents in 1 dollar.

The following coins and bills are used on Beast Island:

Coins

 Penny: 1 cent

 Nickel: 5 cents

 Dime: 10 cents

 Quarter: 25 cents

Bills

 1-dollar bill

 5-dollar bill

 10-dollar bill

 20-dollar bill

1 dollar and 25 cents can be written $1.25. The decimal point separates the number of dollars from the number of cents. $1.05 represents 1 dollar and 5 cents.

PRACTICE | Use the information above to help you solve each problem below.

56. Grogg has a penny, a nickel, a dime, and a quarter. What is the value in cents of Grogg's coins?

56. _____

57. Ralph has four nickels and three dimes. Cammie has four dimes and three nickels. How many more cents does Cammie have than Ralph?

57. _____

58. Fiona has 4 bills worth a total of 25 dollars. How many 10-dollar bills does she have?

58. _____

PRACTICE	Use the information on the previous page to help you solve each problem below.

59. Drew has an equal number of quarters and nickels. If Drew has 90 cents in quarters and nickels, how many quarters does he have?

59. _____

60. Max has one dollar bill, two quarters, and three dimes. Lizzie has the same amount of money, all in nickels. How many nickels does Lizzie have?

60. _____

61. ★ Sally has 18 coins. Some are quarters, and the rest are dimes. The amount of money Sally has in dimes is twice the amount she has in quarters. What is the total value in cents of Sally's coins?

61. _____

62. ★ In the Beast Academy cafeteria, Grogg buys a half-pint of milk using only dimes. Lizzie buys a half-pint of milk with the same number of coins, but uses only pennies and quarters. What is the smallest possible price of a half-pint of milk in the Beast Academy cafeteria?

62. _____

63. ★ Grogg has 10 coins worth a total of 32 cents. How many nickels does he have?

63. _____

64. ★ Ms. Q. has 6 bills worth a total of $41. Captain Kraken also has 6 bills worth a total of $41, but he has more $5 bills than Ms. Q. How many $5 bills does Captain Kraken have?

64. _____

The time of day is displayed on a clock.

A clock that uses hands to tell time is called an **analog clock**.

EXAMPLE | What time is displayed on the clock below?

The short hand on the clock tells us the hour. The hour hand is between the 2 and the 3, so the current time is between 2:00 and 3:00.

The long hand tells us the number of minutes past the hour. Each of the small tick marks stands for one minute. The tick marks can be numbered starting with 0 on top, and moving around the clock as shown. On this clock, the minute hand is on the 37th tick mark, which tells us the time is 37 minutes past the hour.

Time is written with the hour and minute separated by a colon. The time displayed on the clock is **2:37**.

When writing the time, the abbreviation **a.m.** means between midnight and noon (midday), and **p.m.** means between noon and midnight. The clock above does not tell us whether the time is 2:37 a.m. or 2:37 p.m.

PRACTICE | Write the time displayed on each clock below.

65.

66.

65. _____

66. _____

PRACTICE | Draw the hour and minute hands to display the time given below each clock.

67.

3:33

68.

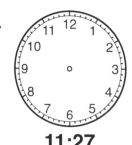

11:27

PRACTICE

69. What time will it be 45 minutes after the time shown below?

69. _____

70. Class ends at 2:15 p.m. The current time is shown below. How many minutes are left in class?

70. _____

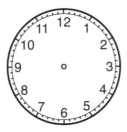

71. A movie starts at 12:25 p.m. and is 1 hour and 43 minutes long. Draw hands on the clock below to indicate the time that the movie ends.

72. It is now 4:56 a.m. Draw hands on the clock below to indicate what time it will be in 10 hours and 10 minutes.

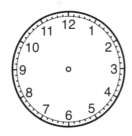

There are many different units of time.

Units of Time

Unit	Conversion
second (sec)	
minute (min)	1 min = 60 sec
hour (hr)	1 hr = 60 min
day	1 day = 24 hr
week	1 week = 7 days
month	Each month has 28, 29, 30, or 31 days
year	1 year has 12 months (365* days)
decade	1 decade = 10 years
century	1 century = 100 years
millennium	1 millennium = 1,000 years

*A leap year occurs every 4 years and has 366 days. The year 2000 was a leap year.

PRACTICE | Use the information above to help you solve each problem below.

73. How many hours are there in three days?

73. _____

74. How many seconds are there in one hour?

74. _____

75. What time will it be 20 hours after 5:00 p.m.?

75. _____

76. How many hours pass from Wednesday at 5:00 p.m. to Friday at 9:00 a.m. of the same week?

76. _____

Beast Academy Practice 3C

PRACTICE | Use the information on the previous page to help you solve each problem below.

77. How many decades are there in one millennium?

77. _____

78. How many days are in a month that begins and ends on a Saturday?

78. _____

79. What is the greatest possible number of days that can occur in a decade? (Remember, leap years occur every four years.)

79. _____

80. ★ What time will it be 234 minutes after 2:34 p.m.?

80. _____

81. ★ While Sergeant Rote is sleeping, the power goes out. When the power comes back on, his digital clock resets to 12:00, then runs as usual. When he awakes at 6:05, his alarm clock displays 3:17. What time did the power come back on at Sergeant Rote's house?

81. _____

82. ✏ ★ When asked his age, a little monster said, "The day before yesterday, I was 6. Next year, I turn 9." How is this possible?

Compute your age in days:

How many days old are you? Use the information below to help you find out!

Below are the number of days in each month.

Jan	Feb	Mar	April	May	June	July	Aug	Sept	Oct	Nov	Dec
31	28*	31	30	31	30	31	31	30	31	30	31

*In a leap year, February has 29 days.

To compute your age in days, it helps to know how many days you lived during each calendar year since you were born.

EXAMPLE | Addie was born on October 12, 2004. How many days old was Addie on March 23, 2012?

We can organize our work with a chart like the one on the right.

Since she was born on October 12, Addie lived for 31–12=19 full days in October of 2004. She also lived for 30 days in November and 31 days in December of 2004, for a total of 19+30+31=80 days in 2004.

Addie lived for 7 complete calendar years (2005-2011). 2008 was a leap year. So, Addie lived for 366 days in 2008, and 365 days for each of the other full years.

Year	Days Lived
2004	19+30+31=80
2005	365
2006	365
2007	365
2008	366
2009	365
2010	365
2011	365
2012	31+29+23=83

Next, we compute the number of days Addie lived in 2012. Addie lived through all of January (31 days), all of February (29 days, since 2012 is a leap year), and 23 days in March, for a total of 31+29+23=83 days.

Finally, we add the totals for each year.
We can use multiplication to add the seven full years:

　 365+365+365+366+365+365+365
=365+365+365+365+1+365+365+365
=(7×365)+1=2,556.

Then, add the days Addie lived in 2004 and 2012 to get 2,556+80+83=2,719.

Addie was **2,719** days old on March 23, 2012.

How many days old are you today?

EXAMPLE | Complete the sentence below with the best answer from the choices provided.

Choosing the right unit can make measurements easier to understand.

The amount of time you sleep each night is best described in _____.
(seconds, hours, or weeks)

The amount you sleep each night is much less than one week. In a whole month, most people only sleep a little more than 1 week all together. The number of seconds most people sleep each night is very large.

Most people sleep between 6 and 10 hours each night.

You can describe how long you sleep each night in any unit of time, but hours is easiest to understand, because the number of hours is not very large or very small. So, the amount of time you sleep each night is best described in **hours**.

PRACTICE | Complete each sentence below with the best answer from the choices provided.

83. The time it takes to microwave popcorn is best described in _____.
    (seconds or hours)

84. The weight of one orange is best described in _____.
    (ounces or tons)

85. The volume of soda in a can is best described in _____.
    (fluid ounces or gallons)

86. The height of a house is best described in _____.
    (inches or feet)

87. The capacity of a trash can is best described in _____.
    (ounces or gallons)

88. The age of a large tree is best described in _____.
    (hours, weeks, or years)

89. The width of this book is best described in _____.
    (centimeters, meters, or kilometers)

90. The weight of a jumbo jet is best described in _____.
    (ounces, pounds, or tons)

PRACTICE | Complete each sentence below with the best answer from the choices below. Choices include customary and metric units. No answer choice is used more than once, and some choices will not be used at all.

Length: centimeters (cm), inches (in), feet (ft), meters (m), miles (mi)
Weight: grams (g), ounces (oz), pounds (lb), tons
Volume: milliliters (mL), fluid ounces (fl oz), liters (L), gallons (gal)
Time: seconds (sec), minutes (min), hours (hr), months, years

91. An adult caterpillar is about 5 _____ long.

92. My cup contains 11 _____ of juice.

93. I can ride my bike 10 _____ in 1 hour.

94. An eyedropper holds about 2 _____ of liquid.

95. The SUV weighs about 3,500 _____.

96. The height of the kitchen counter is 34 _____.

97. The football game on television was 3 _____ long.

98. Together, a nickel and two pennies weigh about 10 _____.

99. An adult elephant weighs about 5 _____.

100. It takes about 3 _____ to read this sentence.

101. A toilet uses about 10 _____ of water per flush.

| **PRACTICE** | Make your best guess for each value below. Then, find the actual value by measuring it, reading its label, or looking it up. |

102. The length of a dollar bill in inches.

102. _____

103. The number of miles from Los Angeles to New York.

103. _____

104. The weight in ounces of a box of cereal.

104. _____

105. The length in centimeters of this sentence.

105. _____

106. The weight of a quarter in grams.

106. _____

107. The volume of shampoo in a bottle in fluid ounces.

107. _____

108. The weight in pounds of one gallon of milk.

108. _____

109. The capacity in gallons of a kitchen sink.

109. _____

| **PRACTICE** | It is useful to understand approximate relationships between customary and metric units. Fill in each blank below with a customary unit that completes each sentence. |

110. Five kilometers is about the same distance as 3 _____.

111. One liter is a little more than 1 _____.

112. One kilogram is a little more than 2 _____.

113. Two and a half centimeters is about the same length as 1 _____.

114. One meter is a little longer than 1 _____.

PRACTICE

115. For a quarter, Pat can play a video game for 5 minutes. How many dollars worth of quarters does Pat need to play the same video game for an hour?

115. _____

116. Dave's best marathon time is 5 hours and 18 minutes. Patrick's best marathon time is half as long as Dave's. What is Patrick's best marathon time in hours and minutes?

116. ___hr ____min

117. If 3 cups and 7 fluid ounces of water are added to a jug, it will contain 7 cups and 3 fluid ounces of water. How many cups and fluid ounces of water does the jug contain now?

117. ___c ___fl oz

118. The number of inches in 8 feet is equal to the number of fluid ounces in ___ cups.

118. _____

119. Gary is a tiny beast who can carry a pack that is four times his weight. Together, Gary and his pack weigh 1 lb 4 oz. What is Gary's weight in ounces?

119. _____

120. To convert a temperature in degrees Celsius to degrees Fahrenheit, multiply the number of Celsius degrees by 9, divide the result by 5, then add 32. Convert 40°C to degrees Fahrenheit.

120. _____

PRACTICE

121. Grogg's snail can slime one inch every second. How many **minutes**
★ will it take for Grogg's snail to slime 5 yards?

121. _____

122. Four hundred pennies weigh 1 kilogram. Yerg's penny collection
★ weighs 700 grams. How many pennies does Yerg have in his
collection?

122. _____

Use the following information for problems 123-124:
Klurg has two clocks on his night stand: a digital clock and an analog
clock. When the power goes out, both clocks stop running. When the
power comes back on, his digital clock resets to 12:00. His analog
clock continues running from the previous time.

123. One night, while Klurg is asleep, the power turns off at 3:00 and stays
★ out for 45 minutes. What time will be displayed on each of Klurg's
clocks at exactly 7:00?

124. One morning, Klurg wakes up to find his clocks displaying the times
★ shown below. What time was it when the power went out?
★

124. _____

PRACTICE | Professor Grok has only the four weights shown on the right (one of each). Show how some or all of the weights can be placed on the **right side** of each scale below to balance the scale.

125.

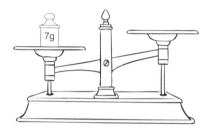

126.

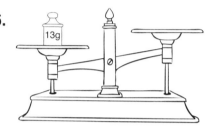

127. Captain Kraken places a gold coin on a balance scale. Professor Grok cannot balance the scale using only the four weights shown above. What is the smallest possible number of grams Kraken's coin could weigh?

127. _____

Use the following for Problems 128 to 130:
Professor Grok has only the four weights shown (one of each). He can place these weights on **both sides** of the scale. For example, Professor Grok can balance the scale below by placing the 27-gram weight on the right, and the 1-gram weight on the left.

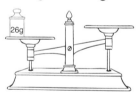

Show how some or all of the four weights above can be placed to balance each scale below.

128.
★

129.
★

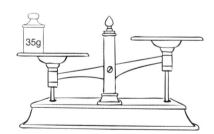

130.
★
★
Grogg places a tribble on a balance scale. Professor Grok cannot balance the scale using only the four weights above. What is the smallest number of grams Grogg's tribble could weigh?

130. _____

Investigations

If you completed the previous projects, you already know that a bathtub can be used to measure volume, and a weighted string can be used to measure time.

In the problems below, you will explore how to use a ruler to measure weight, a stopwatch to measure distance, and a gallon jug to measure time.

Use what you know about units and measures to complete the investigations below.

PRACTICE | For each investigation below, describe one way you could use the tool listed to complete the given task.

131. How could you use a ruler to compare the weight of a dollar worth of dimes to the weight of a dollar worth of quarters?

★

132. How could you use a stopwatch to figure out how far away you are from a lightning strike?

★

133. How could you use a one-gallon milk jug to create a timer that will accurately measure 5 minutes?

★

HINTS
For Selected Problems

Below are hints to every problem marked with a ★.
Work on the problems for a while before looking at the hints.
The hint numbers match the problem numbers.

CHAPTER 7
Variables

36. Start with 13+*g*. What are you left with after subtracting 13 from this expression? Then, what are you left with after subtracting *g* from that?

37. Start with *t*. How many times is *t* added? How many times is *t* subtracted?

41. Grogg starts with 1 toothpick and adds 3 toothpicks to make each new square.

44. Grogg starts with 1 toothpick and adds 2 toothpicks to make each new triangle.

45. Grogg starts with 1 toothpick and adds 4 toothpicks to make each new pentagon.

99. If Devin is *d* years old today, how old was he 15 years ago?

103. After Winnie enters on floor *f* and goes up 6 floors, she is on floor *f*+6.

109. Lizzie read *w* pages on Wednesday. What expression can you write for the number of pages she read on Thursday? Friday?

117. We can label any blank circle whose lower circles are filled in. One example is shown below.

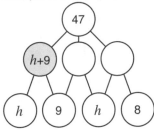
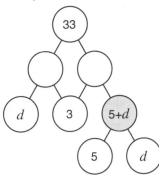

118. We can label any blank circle whose lower circles are filled in. One example is shown below.

119. What expression could you use to label the shaded circle below?

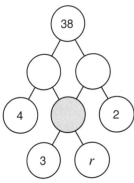

120. What expressions could be used to label the shaded circles below?

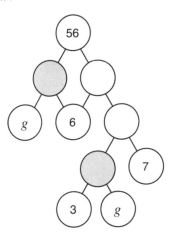

121. What expression could you use to label the shaded circle below?

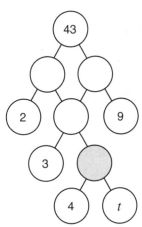

124. We can remove one c-gram weight from each side of the left scale without affecting the balance of the scale.

125. On the right scale, two g-gram weights balance one h-gram weight. So, we can replace any h-gram weight on a scale with two g-gram weights without affecting the balance of the scale.

126. On the left scale, three w-gram weights balance two v-gram weights. So, we can replace any two v-gram weights on a scale with three w-gram weights without affecting the balance of the scale.

130. We can subtract r from both sides of the second equation.

131. Find two expressions that equal $n+4$.

CHAPTER 8
Division   36

62. To fill in the center square, look for a number that divides both 35 and 40.

63. Find the unknown number in the vertical multiplication equation $3\times\square=30$. Then, to fill in the center square, we look for a number that divides the numbers in these gray squares:

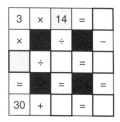

72. Be careful with the zeroes in this problem. Check your division with multiplication.

73. Be careful with the zeroes in this problem. Check your division with multiplication.

84. 14×10=140, so 14 goes into 86 fewer than 10 times. Try a smaller number.

85. 8×300=2,400.

86. What would be the value of n if the remainder when $n\div7$ were 0 instead of 6?

87. The quotient of $77\div m$ is 9 and the remainder is 5. This means that m goes into 77 nine times, with 5 left over. What number could m go into with no remainder?

91. After finding all of the remainders, work backwards. Find a hexagon that can be used to reach the hexagon marked "Finish," and work your way back to the hexagon marked "Start."
Use the hint above for problems 91 through 96.

96. Use the remainders you know to find the remainders you have not found.

108. You don't need to evaluate 151×239 to solve this problem.

109. 64÷1=64. What is 64÷64?
64÷2=32. What is 64÷32?
64÷4=16. What is 64÷16?

110. Replace a, b, and c with values so that $a\div b$ equals c. For example, what is c when a=42 and b=6? Then, compute $a\div c$. What do you notice? Does this always work? Why?

117. Use the remainder of 187÷75 to find the remainder of 188÷75 and 189÷75.

124. You do not need to compute 103×104×105 to find the remainder.

144. Which pair of missing numbers cannot be placed in connected circles? Which of the blank circles could those numbers go in?

145. Which pair of missing numbers cannot be placed in connected circles? Which of the blank circles could those numbers go in?

146. Which pair of missing numbers cannot be placed in connected circles? Which of the blank circles could those numbers go in?

158. Which will Alex run out of first, chocolate chips or cinnamon candies?

159. Don't leave any of the 223 little monsters behind!

160. Begin by finding the numbers that a and b represent.

161. There are 7 days in a week. How many full weeks are there in a leap year? How many days are left over?

162. How many times did Grogg write his name?

CHAPTER 9
Units and Measurement   70

11. How many 40-centimeter towers must be stacked to make a tower that is 2 meters tall?

12. Start with a diagram:

13. How many inches does each new cup add to the height of the stack?

14. Start by finding the longest distance. Then, try converting the other three distances to the same unit to compare them.

22. The answer is less than 17 cm. It may help to label the distances on the diagram between E and F, F and G, G and H, H and J, and J and K.

23. Begin at Gil's and end at Hank's.

37. Add the pounds and ounces separately.

43. Find the total weight of all five items. For the scale to balance, how much weight must be placed on each side?

44. What is weighed by the top scale? What is weighed by the bottom scale?

61. Sally's dimes are worth twice as much as her quarters. How many dimes does Sally have for each quarter she has?

62. What is the smallest number of pennies Lizzie can have?

63. How many pennies does Grogg have?

64. Can you find all of the ways to make $41 with bills worth $20, $10, $5, and $1? Use a chart like the one below to organize your work. The first row is completed for you.

$20	$10	$5	$1	Total number of bills:
2	0	0	1	3

80. How many minutes are there in 4 hours?

81. How much time passed between the time the power came back on and the time Sergeant Rote woke up?

82. On what day was the little monster asked his age? On what day is the little monster's birthday?

121. How many inches are there in 5 yards?

122. How many pennies weigh 100 grams?

123. At what time did the power come back on? How far behind is each clock when the power comes on?

124. Hint #1: When the power came back on, which clock displayed the time that the power went out?
Hint #2: Which clock tells you how long it has been since the power came back on?

128. Which weight must be placed on the right side?

129. Which weight must be placed on the right side?

130. Hint #1: Use only the 1, 3, and 9-gram weights to balance each weight from 1 to 13 grams. For example, to balance 8 grams, you could place the 1-gram weight on the same side as the 8-gram weight, and the 9-gram weight on the other side.
Hint #2: To balance weights greater than 13 grams, start by placing the 27-gram weight on the opposite side.

131. How could you make the ruler work like a balance scale?

132. Have you ever noticed that you see lightning strike before you hear the resulting thunder, and see fireworks explode before hearing them?

133. Start with a jug that is full of water.

SOLUTIONS
Chapters 7-9

VARIABLES
Basics
page 7

1. $9+\boxed{7}=16$.
2. $35-\boxed{15}=20$.
3. $80+20=\boxed{100}$.
4. $\boxed{40}-9=31$.

We can think about the "n" in each of the next four problems like the blank boxes in the previous four!

5. $30+\boxed{100}=130$, so $n=\textbf{100}$.
6. $\boxed{30}+5=35$, so $n=\textbf{30}$.
7. $\boxed{70}-3=67$, so $n=\textbf{70}$.
8. $100-\boxed{60}=40$, so $n=\textbf{60}$.

Evaluating Expressions
8-9

9. When $n=6$, the expression $13+n$ is equal to $13+6=\textbf{19}$.
10. When $n=6$, the expression $33-n$ is equal to $33-6=\textbf{27}$.
11. When $n=6$, the expression $n\times4$ is equal to $6\times4=\textbf{24}$.
12. When $n=6$, the expression $2\times4+n$ is equal to $2\times4+6=8+6=\textbf{14}$.
13. When $r=10$, the expression $r+216$ is equal to $10+216=\textbf{226}$.
14. When $r=10$, the expression $152+9-r$ is equal to $152+9-10=\textbf{151}$.
15. When $r=10$, the expression $122\times r$ is equal to $122\times10=\textbf{1,220}$.
16. When $r=10$, the expression $7\times(r+20)$ is equal to $7\times(10+20)=7\times30=\textbf{210}$.
17. When $a=9$, the expression $9\times a+3$ is equal to $9\times9+3=81+3=\textbf{84}$.
18. When $a=4$, the expression $9\times a+3$ is equal to $9\times4+3=36+3=\textbf{39}$.
19. When $a=20$, the expression $9\times a+3$ is equal to $9\times20+3=180+3=\textbf{183}$.
20. When $k=25$, the expression $300-k\times2$ is equal to $300-25\times2=300-50=\textbf{250}$.
21. When $k=100$, the expression $300-k\times2$ is equal to $300-100\times2=300-200=\textbf{100}$.
22. When $k=60$, the expression $300-k\times2$ is equal to $300-60\times2=300-120=\textbf{180}$.
23. When $d=6$, the expression $3\times(d+4)$ is equal to $3\times(6+4)=3\times10=\textbf{30}$.
24. When $d=10$, the expression $3\times(d+4)$ is equal to $3\times(10+4)=3\times14=\textbf{42}$.
25. When $d=17$, the expression $3\times(d+4)$ is equal to $3\times(17+4)=3\times21=\textbf{63}$.

VARIABLES
Simplifying Expressions
10-11

26. Adding six n's is the same as multiplying 6 times n, so we can write $n+n+n+n+n+n$ as $\textbf{6}\times\textbf{n}$ (or $\textbf{n}\times\textbf{6}$).
27. When you subtract a number from itself, you always get 0. So, $n-n$ simplifies to $\textbf{0}$.
28. Adding 5 and then subtracting 5 is the same as doing nothing. So, $n+5-5$ simplifies to $\textbf{n}$.
29. Adding n and then subtracting n is the same as doing nothing. So, $17+n-n$ simplifies to $\textbf{17}$.
30. Since $11+14=25$, we have $d+11+14=\textbf{d+25}$ (or $\textbf{25+d}$).
31. Because addition is associative and commutative, we can add the numbers and variables in an addition expression in any order. So, $17+p+2=p+17+2$ (or $17+2+p$). Since $17+2=19$, we have $p+17+2=\textbf{p+19}$ (or $\textbf{19+p}$).
32. Since $13-12=1$, we have $13-12+f=1+f$ (or $\textbf{f+1}$).
33. Adding k and then subtracting k is the same as doing nothing. So, $5+k-k=\textbf{5}$.
34. Subtracting 20 and then adding 20 is the same as doing nothing. So, $j-20+20$ simplifies to $\textbf{j}$.
35. We can use the associative and commutative properties of addition to rewrite this expression:
$w+20-w=(w+20)-w=(20+w)-w=20+w-w$.
Adding w and then subtracting w is the same as doing nothing. So, $w+20-w=\textbf{20}$.
36. We start with $13+g$. Subtracting 13 from this expression leaves us with g. Then, subtracting g leaves us with $\textbf{0}$.
37. We start with $t+t$. Subtracting one t leaves us with t. Then, adding two t's and subtracting two t's is the same as doing nothing. So, this expression simplifies to $\textbf{t}$.

— or —

Start with t, we add 3 t's and subtract 3 t's. This leaves us with one $\textbf{t}$.

VARIABLES
Describing Patterns
12-15

38. For every square in the diagram, Grogg needs 4 toothpicks. So, we can multiply the number of squares by 4 to get the number of toothpicks:

Squares	Toothpicks	
1	4	$1\times4=4$.
2	8	$2\times4=8$.
3	12	$3\times4=12$.
4	**16**	$4\times4=16$.
5	**20**	$5\times4=20$.
10	**40**	$10\times4=40$.
100	**400**	$100\times4=400$.

39. For every square in the diagram, Grogg needs 4 toothpicks. So, to make a pattern with n squares, Grogg needs $n\times4$ toothpicks.

Squares	Toothpicks
n	$n\times4$

$n+3$ $n\times3$ $\boxed{n\times4}$ $1+n\times2$ $1+n\times3$

We can check this answer by evaluating the expression $n\times4$ for some values of n in the table. For example, when $n=1$, $n\times4=1\times4=4$. When $n=2$, $n\times4=2\times4=8$. When $n=3$, $n\times4=3\times4=12$. When $n=20$, $n\times4=20\times4=80$. All of these numbers match the values in the table for Problem 38.

40. We evaluate the expression we found in the previous problem for $n=40$. When $n=40$, the expression $n\times4$ is equal to $40\times4=$**160**.

41. To make the first square in the diagram, Grogg needs 4 toothpicks. To make each additional square, he must add 3 more toothpicks. We could also think of Grogg starting with 1 toothpick and adding 3 toothpicks for each square.

Squares	Toothpicks	
1	4	$1+3=4$.
2	7	$1+3+3=1+2\times3=1+6=7$.
3	10	$1+3+3+3=1+3\times3=1+9=10$.
4	**13**	$1+4\times3=1+12=13$.
5	**16**	$1+5\times3=1+15=16$.
20	**61**	$1+20\times3=1+60=61$.
100	**301**	$1+100\times3=1+300=301$.

42. We can think of Grogg starting with 1 toothpick and adding 3 toothpicks for each square. So, if Grogg wants to make a pattern with n squares, he will need $1+n\times3$ toothpicks.

Squares	Toothpicks
n	$1+n\times3$

$n\times5$ $n\times3$ $n+3$ $1+n\times2$ $\boxed{1+n\times3}$

We can check this answer by evaluating the expression $1+n\times3$ for some values of n in the table for Problem 41.

43. We evaluate the expression we found in the previous problem for $n=50$. When $n=50$, the expression $1+n\times3$ is equal to $1+50\times3=1+150=$**151**.

44. To make the first triangle in the diagram, Grogg needs 3 toothpicks. To make each additional triangle, he must add 2 more toothpicks. We could also think of Grogg starting with 1 toothpick and adding 2 toothpicks for each triangle.

Triangles	Toothpicks	
1	3	$1+2=3$.
2	5	$1+2+2=1+2\times2=1+4=5$.
3	7	$1+2+2+2=1+3\times2=1+6=7$.
4	**9**	$1+4\times2=1+8=9$.
5	**11**	$1+5\times2=1+10=11$.
n	**$1+n\times2$**	or **$1+2\times n$**, **$2\times n+1$**, **$n\times2+1$**
100	**201**	$1+100\times2=1+200=201$.

We can check this answer by evaluating the expression $1+n\times2$ for some values of n in the table.

45. To make the first pentagon in the diagram, Grogg needs 5 toothpicks. To make each additional pentagon, he must add 4 more toothpicks. We could also think of Grogg starting with 1 toothpick and adding 4 toothpicks for each pentagon.

Pentagons	Toothpicks	
1	5	$1+4=5$.
2	9	$1+4+4=1+2\times4=1+8=9$.
3	13	$1+4+4+4=1+3\times4=1+12=13$.
4	**17**	$1+4\times4=1+16=17$.
5	**21**	$1+5\times4=1+20=21$.
n	**$1+n\times4$**	or **$1+4\times n$**, **$n\times4+1$**, **$4\times n+1$**
100	**401**	$1+100\times4=1+400=401$.

We can check this answer by evaluating the expression $1+n\times4$ for some values of n in the table.

VARIABLES
Expressions in Geometry   16-17

46. The four sides of a square are all the same length. To find the perimeter of a square, we add the side length 4 times. So, the perimeter of a square with sides of length s is $s+s+s+s$. This can be simplified to **$4\times s$** or **$s\times4$**.

47. The six sides of a regular hexagon are all the same length. To find the perimeter of a regular hexagon, we add the side length 6 times. So, the perimeter of a regular hexagon with sides of length h is $h+h+h+h+h+h$. This can be simplified to **$6\times h$** or **$h\times6$**.

48. A square is a special type of rectangle. To find the area of any rectangle, we can multiply its height by its width. The side length of a square is both its height and its width. So, an expression for the area of a square with side length s is **$s\times s$**.

49. To find the area of a rectangle, we can multiply its height by its width. So, an expression for the area of a rectangle with height 5 and width w is **$5\times w$** (or **$w\times5$**).

50. To find the area of a rectangle, we can multiply its height by its width. So, an expression for the area of a rectangle with height h and width w is **$h\times w$** (or **$w\times h$**).

51. The perimeter of a polygon is the sum of the lengths of all its sides. So, the perimeter of a rectangle with height h and width 20 is **$h+20+h+20$**, which is equal to **$h+h+40$**. You may have also written this as **$h+40+h$**, **$40+h+h$**, **$2\times h+40$**, **$h\times2+40$**, **$40+2\times h$**, or **$40+h\times2$**.

— *or* —

To find the perimeter of a rectangle, we can add its height and width, then double the sum. So, the perimeter of a rectangle with height h and width 20 is **$2\times(h+20)$**. You may have also written this as **$2\times(20+h)$**, **$(20+h)\times2$**, or **$(h+20)\times2$**.

Notice that when we distribute the 2 in one of these expressions, we get an expression from the previous solution. For example, $2\times(h+20)=2\times h+2\times20=2\times h+40$.

52. To find the perimeter of a rectangle, we add the lengths of its sides. So, the perimeter of a rectangle with height h and width w is $h+h+w+w$. Since addition is commutative, you may have added the two h's and the two w's in any order. You may have also written this as $2\times h+2\times w$, $2\times w+2\times h$, $h\times2+w\times2$, or $w\times2+h\times2$.

— or —

To find the perimeter of a rectangle, we can add its height and width, then double the sum. So, the perimeter is $2\times(h+w)$. You may have also written this as $2\times(w+h)$, $(w+h)\times2$, or $(h+w)\times2$.

Notice that when we distribute the 2 in one of these expressions, we get an expression from the previous solution. For example, $2\times(h+w)=2\times h+2\times w=h+h+w+w$.

VARIABLES
Hidden Messages   18–19

53. $2\times\boxed{9}=18$, so $a=$**9**.

54. $6+\boxed{6}=12$, so $b=$**6**.

55. $2+3\times2=2+6=\boxed{8}$, so $d=$**8**.

56. $\boxed{5}\times9=45$, so $e=$**5**.

57. $27=9\times\boxed{3}$, so $h=$**3**.

58. $40=57-\boxed{17}$, so $m=$**17**.

59. $80=93-\boxed{13}$, so $o=$**13**.
Because the letter "o" can be mistaken for the digit 0, we usually avoid using the letter "o" as a variable.

60. $4\times\boxed{7}=28$, so $r=$**7**.

61. $70=\boxed{10}\times7$, so $s=$**10**.

62. $7=\boxed{22}-15$, so $t=$**22**.

63. $2\times10=20$, so $2\times10-w=20-w$.
$20-\boxed{1}=19$, so $w=$**1**.

64. $68+\boxed{11}=79$, so $z=$**11**.

Using our answers to Problems 53-64, we decode Calamitous Clod's riddle:

$$\underset{1}{W}\underset{3}{h}\underset{5}{e}\underset{7}{r}\underset{5}{e}\quad\underset{8}{d}\underset{13}{o}\quad\underset{17}{m}\underset{9}{a}\underset{22}{t}\underset{3}{h}$$

$$\underset{6}{b}\underset{5}{e}\underset{9}{a}\underset{10}{s}\underset{22}{t}\underset{10}{s}\quad\underset{5}{e}\underset{9}{a}\underset{22}{t}\ ?$$

65. $90-\boxed{11}=79$, so $a=$**11**.

66. $5\times5-2=25-2=\boxed{23}$, so $b=$**23**.

67. $\boxed{17}+3=20$, so $d=$**17**.

68. $36-\boxed{12}=24$, so $e=$**12**.

69. $\boxed{16}-12=4$, so $g=$**16**.

70. $150-\boxed{20}=130$, so $i=$**20**.

71. $5\times\boxed{4}=20$, so $j=$**4**.

72. $11\times\boxed{2}=22$, so $l=$**2**.

73. $100=\boxed{5}\times20$, so $m=$**5**.

74. $\boxed{25}+50=75$, so $n=$**25**.

75. $2\times9+6=18+6=\boxed{24}$, so $s=$**24**.

76. $\boxed{18}+4=22$, so $t=$**18**.

Using our answers to Problems 65-76, we decode the answer Calamitous Clod's riddle.

$$\underset{11}{A}\underset{18}{t}\quad\underset{11}{a}\quad\underset{18}{t}\underset{20}{i}\underset{5}{m}\underset{12}{e}\underset{24}{s}$$

$$\underset{18}{t}\underset{11}{a}\underset{23}{b}\underset{2}{l}\underset{12}{e}\ !$$

VARIABLES
Region Sums   20–21

77. First, we use the region with a sum of 18 to write an equation: $6+2+4+\bigcirc=18$. We simplify the expression on the left by adding $6+2+4$. This gives us $12+\bigcirc=18$. Since $12+\langle6\rangle=18$, we place a 6 in the blank hexagon.

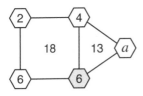

Then, we use the region with a sum of 13 to write an equation: $6+4+a=13$. We simplify the expression on the left by adding $6+4$. This gives us $10+a=13$. Since $10+\underline{3}=13$, $a=$**3**. We replace a with 3 and check our work:

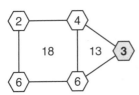

78. First, we use the region with a sum of 12 to write an equation: $6+2+\bigcirc=12$. We simplify the expression on the left by adding $6+2$. This gives us $8+\bigcirc=12$. Since $8+\langle4\rangle=12$, we place a 4 in the center hexagon.

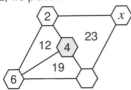

Next, we use the region with a sum of 19 to write an equation: $6+4+\bigcirc=19$. We simplify the expression on the left by adding $6+4$. This gives us $10+\bigcirc=19$. Since $10+\langle9\rangle=19$, we place a 9 in the bottom right hexagon.

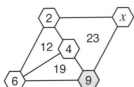

Then, we use the region with a sum of 23 to write an equation: $2+4+9+x=23$. We simplify the expression on the left by adding $2+4+9$. This gives us $15+x=23$. Since $15+\underline{8}=23$, $x=$**8**. We replace x with 8 and check our work:

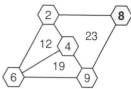

79. First, we use the region with a sum of 18 to write an equation: 5+9+◯=18. We simplify the expression on the left by adding 5+9. This gives us 14+◯=18. Since 14+⟨4⟩=18, we place a 4 in the center hexagon.

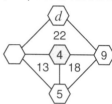

Next, we use the region with a sum of 13 to write an equation: 4+5+◯=13. We simplify the expression on the left by adding 4+5. This gives us 9+◯=13. Since 9+⟨4⟩=13, we place a 4 in the left hexagon.

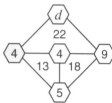

Then, we use the region with a sum of 22 to write an equation: 4+4+9+*d*=22. We simplify the expression on the left by adding 4+4+9. This gives us 17+*d*=22. Since 17+5=22, *d*=**5**. We replace *d* with 5 and check our work:

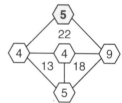

80. First, we use the region with a sum of 11 to write an equation: 6+2+◯=11. We simplify the expression on the left by adding 6+2. This gives us 8+◯=11. Since 8+⟨3⟩=11, we place a 3 in the top right hexagon.

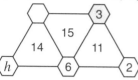

Next, we use the region with a sum of 15 to write an equation: 3+6+◯=15. We simplify the expression on the left by adding 3+6. This gives us 9+◯=15. Since 9+⟨6⟩=15, we place a 6 in the top left hexagon.

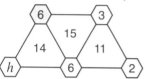

Then, we use the region with a sum of 14 to write an equation: 6+6+*h*=14. We simplify the expression on the left by adding 6+6. This gives us 12+*h*=14. Since 12+2=14, *h*=**2**. We replace *h* with 2 and check our work:

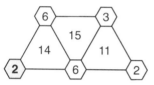

81. First, we use the region with a sum of 16 to write an equation: 7+4+◯=16. We simplify the expression on the left by adding 7+4. This gives us 11+◯=16. Since 11+⟨5⟩=16, we place a 5 in the bottom left hexagon.

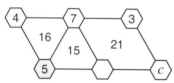

Next, we use the region with a sum of 15 to write an equation: 5+7+◯=15. We simplify the expression on the left by adding 5+7. This gives us 12+◯=15. Since 12+⟨3⟩=15, we place a 3 in the bottom center hexagon.

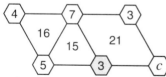

Then, we use the region with a sum of 21 to write an equation: 3+7+3+*c*=21. We simplify the expression on the left by adding 3+7+3. This gives us 13+*c*=21. Since 13+8=21, *c*=**8**. We replace *c* with 8 and check our work:

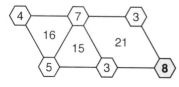

82. First, we use the region with a sum of 35 to write an equation: $14+3+\bigcirc=35$. We simplify the expression on the left by adding $14+3$. This gives us $17+\bigcirc=35$. Since $17+\langle18\rangle=35$, we place an 18 in the top hexagon.

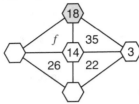

Next, we use the region with a sum of 22 to write an equation: $14+3+\bigcirc=22$. We simplify the expression on the left by adding $14+3$. This gives us $17+\bigcirc=22$. Since $17+\langle5\rangle=22$, we place a 5 in the bottom hexagon.

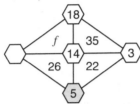

Next, we use the region with a sum of 26 to write an equation: $5+14+\bigcirc=26$. We simplify the expression on the left by adding $5+14$. This gives us $19+\bigcirc=26$. Since $19+\langle7\rangle=26$, we place a 7 in the left hexagon.

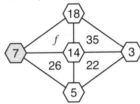

Finally, we use the region with a sum of f to write an equation: $7+14+18=f$. We simplify the expression on the left by adding $7+14+18$. This gives us **39**$=f$. We replace f with 39 and check our work:

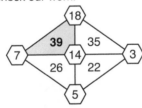

VARIABLES
Balance Scales   22–23

83. The a-gram weight balances a 5-gram weight and a 23-gram weight. We write an equation for the scale: $a=5+23$. Adding $5+23$ gives us $a=28$.

84. The d-gram weight and 3-gram weight are balanced by the 15-gram weight. We write an equation for the scale: $d+3=15$.
If we remove 3 grams from each side of the scale, the d-gram weight will balance $15-3=12$ grams.
Similarly, we solve our equation by subtracting 3 from both sides:

$$\begin{array}{r} d+3=15 \\ -3 \quad -3 \\ \hline d=12 \end{array}$$

So, $d=12$.

85. The 20-gram weight balances the j-gram weight and the 14-gram weight. We write an equation for the scale: $20=j+14$.
If we remove 14 grams from each side of the scale, the j-gram weight will balance $20-14=6$ grams.
Similarly, we solve our equation by subtracting 14 from both sides:

$$\begin{array}{r} 20=j+14 \\ -14 \quad -14 \\ \hline 6=j \end{array}$$

So, $j=6$.

86. The 15-gram weight and the n-gram weight balance the 31-gram weight and the 9-gram weight. We write an equation for the scale: $15+n=31+9$.
This can be simplified to $15+n=40$.
If we remove 15 grams from each side of the scale, the n-gram weight will balance $40-15=25$ grams.
Similarly, we solve our equation by subtracting 15 from both sides:

$$\begin{array}{r} 15+n=40 \\ -15 \quad -15 \\ \hline n=25 \end{array}$$

So, $n=25$.

87. The 53-gram weight and the 8-gram weight balance the h-gram weight and the 13-gram weight. We write an equation for the scale: $53+8=h+13$.
This can be simplified to $61=h+13$.
If we remove 13 grams from each side of the scale, the h-gram weight will balance $61-13=48$ grams.
Similarly, we solve our equation by subtracting 13 from both sides:

$$\begin{array}{r} 61=h+13 \\ -13 \quad -13 \\ \hline 48=h \end{array}$$

So, $h=48$.

88. The 32-gram weight and the w-gram weight balance three 20-gram weights. We write an equation for the scale: $32+w=20+20+20$.
This can be simplified to $32+w=60$.
If we remove 32 grams from each side of the scale, the w-gram weight will balance $60-32=28$ grams.
Similarly, we solve our equation by subtracting 32 from both sides:

$$\begin{array}{r} 32+w=60 \\ -32 \quad -32 \\ \hline w=28 \end{array}$$

So, $w=28$.

VARIABLES
Translating Sentences   24–26

89. "The sum of six and a" means $6+a$ (or $a+6$), and "is" means "equals." So, our equation is $6+a=36$ (or $a+6=36$). To solve the equation, we subtract 6 from both sides:

$$\begin{array}{r} 6+a=36 \\ -6 \quad -6 \\ \hline a=30 \end{array}$$

So, $a=30$.

90. "Twenty-six more than q" means $26+q$ (or $q+26$), and "is" means "equals." So, our equation is $\textbf{26+}\boldsymbol{q}\textbf{=70}$ (or $\boldsymbol{q}\textbf{+26=70}$). To solve the equation, we subtract 26 from both sides:

$$\begin{array}{r} 26+q=70 \\ \underline{-26 \quad\quad -26} \\ q=44 \end{array}$$

So, $q=\textbf{44}$.

91. A "sum" is the result of addition, so "the sum of w and fifteen" means $w+15$ (or $15+w$). The word "is" means "equals," so our equation is $\textbf{43=}\boldsymbol{w}\textbf{+15}$ (or $\textbf{43=15+}\boldsymbol{w}$). To solve the equation, we subtract 15 from both sides.

$$\begin{array}{r} 43=w+15 \\ \underline{-15 \quad\quad -15} \\ 28=w \end{array}$$

So, $w=\textbf{28}$.

92. "Thirty less than t" means $t-30$ (*not* $30-t$), and "is" means "equals." So, our equation is $\boldsymbol{t}\textbf{-30=87}$. To solve the equation, we add 30 to both sides.

$$\begin{array}{r} t-30 = 87 \\ \underline{+30 \quad +30} \\ t=117 \end{array}$$

So, $t=\textbf{117}$.

93. "Nineteen less than j" means $j-19$ (*not* $19-j$), and "is" means equals. So, our equation is $\boldsymbol{j}\textbf{-19=74}$. To solve the equation, we add 19 to both sides.

$$\begin{array}{r} j-19=74 \\ \underline{+19 \quad +19} \\ j=93 \end{array}$$

So, $j=\textbf{93}$.

94. A "sum" is the result of addition, and we are asked to use n to represent "the number." So, "the sum of a number and six" means $n+6$ (or $6+n$).
The word "is" means "equals," so our equation is $\textbf{73=}\boldsymbol{n}\textbf{+6}$ (or $\textbf{73=6+}\boldsymbol{n}$).
To solve the equation, we subtract 6 from both sides.

$$\begin{array}{r} 73=n+6 \\ \underline{-6 \quad\quad -6} \\ 67=n \end{array}$$

So, $n=\textbf{67}$.

95. We are asked to use n to represent "the number," so "twelve less than a number" means $n-12$ (*not* $12-n$). The word "is" means "equals," so our equation is $\boldsymbol{n}\textbf{-12=29}$.
To solve the equation, we add 12 to both sides.

$$\begin{array}{r} n-12 = 29 \\ \underline{+12 \quad +12} \\ n = 41 \end{array}$$

So, $n=\textbf{41}$.

96. We are asked to use g to represent the number of inches in Grogg's height. So, "seven inches more than Grogg's height" means $g+7$ (or $7+g$). Our equation is $\boldsymbol{g}\textbf{+7=65}$ (or $\textbf{7+}\boldsymbol{g}\textbf{=65}$).
To solve the equation, we subtract 7 from both sides.

$$\begin{array}{r} g+7=65 \\ \underline{-7 \quad -7} \\ g=58 \end{array}$$

So, $g=\textbf{58}$.

97. We are asked to use m to represent the number of math books, so "six less than the number of math books" means $m-6$ (*not* $6-m$). Our equation is $\boldsymbol{m}\textbf{-6=15}$. To solve the equation, we add 6 to both sides.

$$\begin{array}{r} m-6=15 \\ \underline{+6 \quad +6} \\ m=21 \end{array}$$

So, $m=\textbf{21}$.

98. We are asked to use p to represent the number of pandakeets, so "fifty-nine more than the number of pandakeets" means $p+59$ (or $59+p$). Our equation is $\textbf{97=}\boldsymbol{p}\textbf{+59}$ (or $\textbf{97=59+}\boldsymbol{p}$).
To solve the equation, we subtract 59 from both sides.

$$\begin{array}{r} 97=p+59 \\ \underline{-59 \quad\quad -59} \\ 38=p \end{array}$$

So, $p=\textbf{38}$.

99. Devin is d years old today. Fifteen years ago, Devin was $d-15$ years old. Our equation is $\boldsymbol{d}\textbf{-15=33}$ (or $\textbf{33=}\boldsymbol{d}\textbf{-15}$).
To solve the equation, we add 15 to both sides.

$$\begin{array}{r} d-15 = 33 \\ \underline{+15 \quad +15} \\ d = 48 \end{array}$$

So, $d=\textbf{48}$.

100. Ralph is r years old today. So, in 37 years, Ralph will be $r+37$ years old. Our equation is $\boldsymbol{r}\textbf{+37=44}$.
To solve the equation, we subtract 37 from both sides.

$$\begin{array}{r} r+37 = 44 \\ \underline{-37 \quad -37} \\ r = 7 \end{array}$$

So, $r=\textbf{7}$.

101. Winnie made w cookies, and Alex made 36 cookies. Together, Winnie and Alex made $w+36$ cookies, so our equation is $\boldsymbol{w}\textbf{+36=84}$.
To solve the equation, we subtract 36 from both sides.

$$\begin{array}{r} w+36 = 84 \\ \underline{-36 \quad -36} \\ w = 48 \end{array}$$

So, $w=\textbf{48}$.

102. There are 65 adult hexatoads and h baby hexatods. All together, there are $65+h$ hexatoads in Grok's office, so our equation is $\textbf{65+}\boldsymbol{h}\textbf{=122}$. To solve the equation, we subtract 65 from both sides.

$$\begin{array}{r} 65+h=122 \\ \underline{-65 \quad\quad -65} \\ h= 57 \end{array}$$

So, $h=\textbf{57}$.

103. If Winnie enters the elevator on floor f and goes up 6 floors, she'll be on floor $f+6$. When she goes down 4 floors, she'll be on floor $f+6-4$. Finally, when she goes up two more floors and exits, she is on floor $f+6-4+2$. We are also told that Winnie exits on floor 9, so our equation is $f+6-4+2=9$. Adding 6, subtracting 4, and then adding 2 is the same as adding $6-4+2=4$, so our equation can be simplified to $f+4=9$.

To solve the equation, we subtract 4 from both sides.

$$\begin{array}{r} f+4 = 9 \\ -4 \; -4 \\ \hline f = 5 \end{array}$$

So, $f=$**5**.

We check our answer: Winnie enters the elevator on floor 5. She goes up 6 floors to floor 11, down 4 floors to floor 7, then up 2 more floors to floor 9, where she exits. ✓

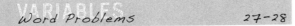

Word Problems   27-28

In each of the problems below, you may have chosen a different letter for your variable. That's okay!

104. We use a to represent the number of gumballs Alex has. Grogg has 12 more gumballs than Alex has, so Grogg has $a+12$ gumballs. All together, they have 32 gumballs, so $a+(a+12)=32$. The associative property of addition lets us remove parentheses from a sum: $a+a+12=32$. Subtracting 12 from both sides, we get $a+a=20$. Since $10+10=20$, we have $a=10$. So, Alex has **10** gumballs.

We check our answer: Alex has 10 gumballs. So, Grogg has $10+12=22$ gumballs. All together, they have $10+22=32$ gumballs. ✓

105. We use r to represent Ralph's height in inches. Olivia is 6 inches taller than Ralph, so Olivia is $r+6$ inches tall. The sum of their heights is 64 inches, so $r+(r+6)=64$. We remove the parentheses to get $r+r+6=64$. Subtracting 6 from both sides, we get $r+r=58$. Since $29+29=58$, we have $r=29$. Ralph is **29** inches tall.

106. We use c to represent the cost in dollars of the case. The xylophone costs $80 more than the case, so the xylophone costs $c+80$ dollars. Together, the xylophone and case cost $100, so $c+(c+80)=100$. We remove the parentheses to get $c+c+80=100$. Subtracting 80 from both sides gives $c+c=20$. Since $10+10=20$, we have $c=10$. So, the case costs **10 dollars ($10)**.

107. We use m to represent the number of dollars Alex earned Monday. Alex earned six dollars more Tuesday than he earned Monday. So, on Tuesday he earned $m+6$ dollars. Alex earned a total of $42, so $m+(m+6)=42$. We remove the parentheses to get $m+m+6=42$. Subtracting 6 from both sides, we get $m+m=36$. Since $18+18=36$, we have $m=18$. Alex earned **18 dollars ($18)** on Monday.

108. We use p to represent the number of points scored by Fiona in the second game. Fiona scored nine fewer points in the first game than she scored in the second game, so she scored $p-9$ points in the first game. All together, she scored 33 points, so $(p-9)+p=33$. We remove the parentheses to get $p-9+p=33$. Adding 9 to both sides, we get $p+p=42$. Since $21+21=42$, we have $p=21$. So, Fiona scored **21** points in the second game.

— *or* —

We use f to represent the number points scored in the *first* game. Fiona scored nine fewer points in the first game than she scored in the second game. So, she scored nine *more* points in the second game than she scored in the first game. So, Fiona scored $f+9$ points in the second game. All together, she scored 33 points, so $f+(f+9)=33$. We remove the parentheses to get $f+f+9=33$. Subtracting 9 from each side, we get $f+f=24$. Since $12+12=24$, we have $f=12$. So, Fiona scored 12 points in the first game and $12+9=$**21** points in the second game.

109. We use w to represent the number of pages Lizzie read on Wednesday. On Friday, she read 7 pages more than she read on Wednesday. So, she read $w+7$ pages on Friday. On Thursday, she read 6 pages fewer than she read on Wednesday. So she read $w-6$ pages on Thursday.
All together, she read 61 pages, so our equation is

$$w+(w+7)+(w-6)=61.$$

We remove the parentheses to get $w+w+7+w-6=61$. Adding 7 and subtracting 6 is the same as adding $7-6=1$, so we simplify the expression on the left above to get the following equation:

$$w+w+w+1=61.$$

Subtracting 1 from both sides, we get $w+w+w=60$. Since $20+20+20=60$, we have $w=20$. So, Lizzie read **20** pages of her book on Wednesday.

Circle Sums   29-31

110. Each circle is labeled with the sum of the numbers below it. So, we use the three circles to write an equation:

$$p+16=58.$$

Subtracting 16 from both sides, we get $p=$**42**.

111. Each circle is labeled with the sum of the numbers below it. So, we label the blank circle $c+16$ as shown.

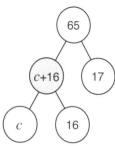

Next, we use the top three circles to write an equation: $(c+16)+17=65$.
We remove parentheses and simplify by adding 16 and 17 to get $c+33=65$.
Subtracting 33 from both sides, we get $c=\textbf{32}$.

112. Each circle is labeled with the sum of the numbers below it. So, we label the left blank circle $18+3=21$ and the right blank circle $3+k$ as shown.

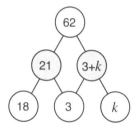

Next, we use the top three circles to write an equation: $21+(3+k)=62$.
We remove parentheses and simplify by adding 21 and 3 to get $24+k=62$.
Subtracting 24 from both sides, we get $k=\textbf{38}$.

113. We label the left blank circle $n+12$ and the right blank circle $12+2=14$, as shown.

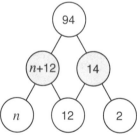

Next, we use the top three circles to write an equation: $(n+12)+14=94$.
We remove parentheses and simplify by adding 12 and 14 to get $n+26=94$.
Subtracting 26 from both sides, we get $n=\textbf{68}$.

114. We label the blank circle $6+a$, as shown.

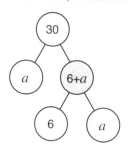

Next, we use the top three circles to write an equation: $a+(6+a)=30$.
We remove parentheses to get $a+6+a=30$.
Subtracting 6 from both sides, we get $a+a=24$.
Since $12+12=24$, we have $a=\textbf{12}$.

115. We label the left blank circle $5+f$, and we label the right blank circle $f+2$.

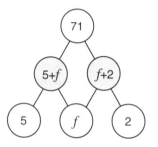

Now, we use the top three circles to write an equation: $(5+f)+(f+2)=71$.
We remove parentheses and simplify by adding 5 and 2 to get $7+f+f=71$.
Subtracting 7 from both sides, we get $f+f=64$.
Since $32+32=64$, we have $f=\textbf{32}$.

116. We label the left blank circle $3+7=10$, and we label the right blank circle $5+m$, as shown.

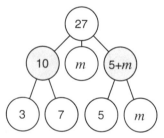

Next, we use the top four circles to write an equation: $10+m+(5+m)=27$.
We remove parentheses and simplify by adding 10 and 5 to get $15+m+m=27$.
Subtracting 15 from both sides, we get $m+m=12$.
Since $6+6=12$, we have $m=\textbf{6}$.

117. We label the left blank circle $h+9$, the middle blank circle $9+h$, and the right blank circle $h+8$ as shown.

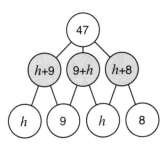

Next, we use the top four circles to write an equation:
$(h+9)+(9+h)+(h+8)=47$.
We remove the parentheses and simplify by adding 9+9+8 to give us $26+h+h+h=47$.
Subtracting 26 from both sides, we get $h+h+h=21$.
Since 7+7+7=21, we have $h=\textbf{7}$.

118. We first label the blank circles whose lower circles are filled in, as shown. This will help us fill in the remaining blank circles.

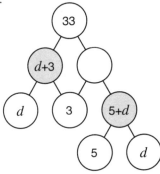

Now, we label the final blank circle with $3+(5+d)$, which simplifies to $8+d$.

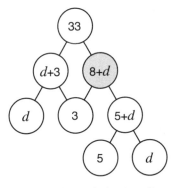

Finally, we use the top three circles to write an equation:
$(d+3)+(8+d)=33$.
We remove parentheses and simplify by adding 3 and 8 to get $11+d+d=33$.
Subtracting 11 from both sides, we get $d+d=22$.
Since 11+11=22, we have $d=\textbf{11}$.

119. We first label the blank circle whose lower circles are filled in, as shown. This will help us fill in the remaining blank circles.

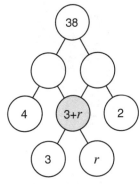

Now, we label the other two blank circles. The blank circle on the left is $4+(3+r)$, which simplifies to $7+r$. The blank circle on the right is $(3+r)+2$, which simplifies to $r+5$.

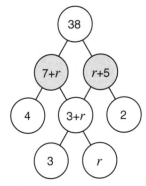

Finally, we use the top three circles to write an equation:
$(7+r)+(r+5)=38$.
We remove parentheses and simplify by adding 7 and 5 to get $12+r+r=38$.
Subtracting 12 from both sides, we get $r+r=26$.
Since 13+13=26, we have $r=\textbf{13}$.

120. We label the blank circles whose lower circles are filled in, as shown. This will help us fill in other blank circles.

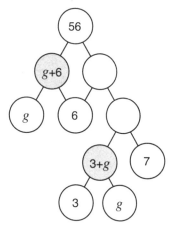

Next, we label the lower blank circle. The lower blank circle is $(3+g)+7$, which simplifies to $10+g$.

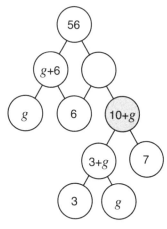

Then, we label the remaining blank circle $6+(10+g)$, which simplifies to $16+g$.

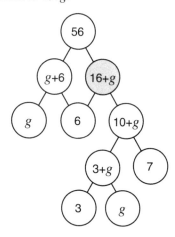

Finally, we use the top three circles to write an equation: $(g+6)+(16+g)=56$.
We remove parentheses and simplify by adding 6 and 16 to get $22+g+g=56$.
Subtracting 22 from both sides, we get $g+g=34$.
Since $17+17=34$, we have $g=\mathbf{17}$.

121. We label the blank circle whose lower circles are filled in, as shown. This will help us fill in other blank circles.

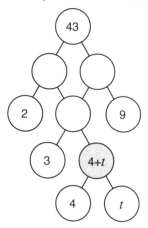

Next, we label one more blank circle. The lower blank circle is $3+(4+t)$, which simplifies to $7+t$.

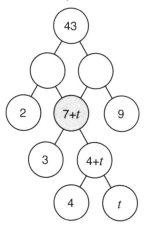

Then, we label the two remaining blank circles. We label the left blank circle $2+(7+t)$, which simplifies to $9+t$. We label the right blank circle $(7+t)+9$, which simplifies to $16+t$.

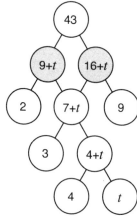

Finally, we use the top three circles to write an equation: $(9+t)+(16+t)=43$.
We remove parentheses and simplify by adding 9 and 16 to get $25+t+t=43$.
Subtracting 25 from both sides, we get $t+t=18$.
Since $9+9=18$, we have $t=\mathbf{9}$.

122. We write an equation to represent each scale:

$$j=12 \quad \text{and} \quad k=j+9.$$

So, $j=\mathbf{12}$.
Since $j=12$, we replace the j-gram weight on the right scale with a 12-gram weight.

This gives us $k=12+9$. We add $12+9$ to get $k=\mathbf{21}$.

We replace j with 12 and k with 21 and check our work:
$$12=12 \checkmark \quad \text{and} \quad 21=12+9 \checkmark$$

123. We write an equation to represent each scale:

$$3+t=9 \quad \text{and} \quad u+u=t.$$

Since the left scale has only one variable, we use the left scale to solve for t first.
If we remove 3 grams from each side of the scale, the t-gram weight will balance $9-3=6$ grams. So, $t=\mathbf{6}$.
Since $t=6$, we replace the t-gram weight on the right scale with a 6-gram weight.

This gives us $u+u=6$. Since $3+3=6$, we have $u=\mathbf{3}$.

We replace t with 6 and u with 3 and check our work:
$$3+6=9 \checkmark \quad \text{and} \quad 3+3=6 \checkmark$$

124. We write an equation to represent each scale:

$$c+c=6+c+d \quad \text{and} \quad d=5+3.$$

Since the right scale has only one variable, we use the right scale to solve for d first.
We add $5+3$ to get $d=\mathbf{8}$.
Since $d=8$, we replace the d-gram weight on the left scale with an 8-gram weight.

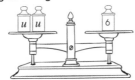

So, the two c-gram weights will balance one 6-gram weight, one c-gram weight, and one 8-gram weight:
$c+c=6+c+8$.

Then, if we remove one c-gram weight from each side of the scale, one c-gram weight will balance a 6-gram weight and an 8-gram weight.

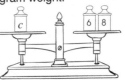

This is the same as subtracting c from both sides of the equation $c+c=6+c+8$. This gives us $c=6+8$.
We add $6+8$ to get $c=\mathbf{14}$.

We replace c with 14 and d with 8 and check our work:
$$14+14=6+14+8 \checkmark \quad \text{and} \quad 8=5+3 \checkmark$$

125. We write an equation to represent each scale:

$$h=g+7 \quad \text{and} \quad h=g+g.$$

On the right scale, two g-gram weights balance one h-gram weight. So, we can replace any h-gram weight on a scale with two g-gram weights without affecting the balance of the scale.
We replace the h-gram weight on the left scale with two g-gram weights.

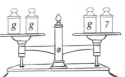

This gives us $g+g=g+7$.
Then, if we remove one g-gram weight from each side of the scale, one g-gram weight will balance 7 grams.

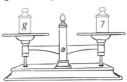

So, $g=\mathbf{7}$.
Since $g=7$, we replace each g-gram weight with a 7-gram weight. This gives us $h=7+7$, so $h=\mathbf{14}$.

We replace g with 7 and h with 14 and check our work:
$$14=7+7 \checkmark \quad \text{and} \quad 14=7+7 \checkmark$$

— *or* —

The only difference between the left scale and the right scale is the pair of shaded weights below. So, the two shaded weights must be equal.

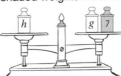 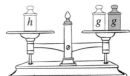

So, $g=\mathbf{7}$.
Since $g=7$, we replace each g-gram weight with a 7-gram weight. This gives us $h=7+7$, so $h=\mathbf{14}$.

126. We write an equation for each scale:
$$v+v=w+w+w \quad \text{and} \quad v+v=w+12.$$

On the left scale, three w-gram weights balance two v-gram weights. So, we can replace any two v-gram weights on a scale with three w-gram weights without affecting the balance of the scale.

We replace the two v-gram weights on the right scale with three w-gram weights.

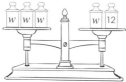

This gives us $w+w+w=w+12$.

Then, if we remove one w-gram weight from both sides of the scale, two w-gram weights will balance 12 grams.

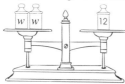

So, $w+w=12$. Since $6+6=12$, $w=$**6**.

Then, since $w=6$, we replace each w-gram weight with a 6-gram weight.

On the left scale, two v-gram weights balance three 6-gram weights: $v+v=6+6+6$. So, $v+v=18$. Since $9+9=18$, we have $v=$**9**.

We replace v with 9 and w with 6 and check our work:
$$9+9=6+6+6 \checkmark \quad \text{and} \quad 9+9=6+12 \checkmark$$
— or —

The only difference between the left scale and the right scale is the set of shaded weights below. So, the 12-gram weight on the left must equal the two w-gram weights on the right.

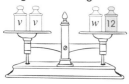 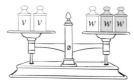

So, $w+w=12$. Since $6+6=12$, we have $w=$**6**.
Since $w=6$, we replace each w-gram weight with a 6-gram weight.
This gives us $v+v=6+12$. So, $v+v=18$.
Since $9+9=18$, we have $v=$**9**.

127. We use the top equation to solve for x:
$$18=x+4.$$

Subtracting 4 from both sides gives us **14**$=x$.
Since $x=14$, we replace the x in the bottom equation with 14 to get $14=y+8$.
Subtracting 8 from both sides, we get $6=y$. So, $y=$**6**.

We replace x with 14 and y with 6 and check our work:
$$18=14+4 \checkmark \quad \text{and} \quad 14=6+8 \checkmark$$

128. Since the bottom equation contains only one variable, we use the bottom equation to solve for j:
$$19=12+j.$$
Subtracting 12 from both sides gives us **7**$=j$.
Since $j=7$, we replace the j in the top equation with 7 to get $7+3=k$.
Adding 7 and 3, we get $10=k$. So, $k=$**10**.

We replace j with 7 and k with 10 and check our work:
$$7+3=10 \checkmark \quad \text{and} \quad 19=12+7 \checkmark$$

129. Since the top equation contains only one variable, we use the top equation to solve for c:
$$c+9=15.$$
Subtracting 9 from both sides gives us $c=$**6**.
Since $c=6$, we replace both c's in the bottom equation with 6's to get $d=6+6$. We add $6+6$ to get $d=$**12**.
We replace c with 6 and d with 12 and check our work:
$$6+9=15 \checkmark \quad \text{and} \quad 12=6+6 \checkmark$$

130. We use the first equation to solve for q:
$$q-6=10.$$

Adding 6 to both sides of the equation gives us $q=$**16**.
Since $q=16$, we replace the q in the first equation with 16 to get $16+r=7+r+r$.
Then, we subtract r from both sides of the equation to get $16=7+r$.
Finally, we subtract 7 from both sides to get $9=r$.
So, $r=$**9**.

We replace q with 16 and r with 9 and check our work:
$$16-6=10. \checkmark \quad \text{and} \quad 16+9=7+9+9. \checkmark$$

131. In the top equation, we see that $m+m$ is equal to $n+4$. In the bottom equation, we see that $16+m$ is also equal to $n+4$. Since $m+m$ and $16+m$ are each equal to $n+4$, we know that $m+m$ and $16+m$ are equal to each other.
$$m+m=16+m.$$
We subtract m from both sides of the equation to get $m=$**16**. Replacing m in both equations with 16, we get
$$16+16=n+4 \quad \text{and} \quad 4+n=16+16.$$
So, $n+4=16+16$. We simplify by adding $16+16$ to get $n+4=32$. Subtracting 4 from both sides gives us $n=$**28**.

We replace m with 16 and n with 28 and check our work:
$$16+16=28+4. \checkmark \quad \text{and} \quad 4+28=16+16. \checkmark$$

DIVISION

Grouping

page 37-38

In each of the following problems, you may have circled different groups of dots to arrive at the same answer.

1. The dots are neatly organized in columns of 3. We circle groups of 3 dots until we run out:

There are **7** groups of 3 dots. 21÷3=7.

2. We circle groups of 4 dots until we run out:

There are **6** groups of 4 dots. 24÷4=6.

3. We circle groups of 8 dots until we run out:

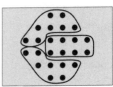

There are **3** groups of 8 dots. 24÷8=3.

4. We circle groups of 9 dots until we run out:

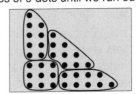

There are **4** groups of 9 dots. 36÷9=4.

5. There are 6 equal rows, each with 5 dots.

So, when the dots are divided into 6 equal groups, **5** dots are in each group. 30÷6=5.

6. The dots are arranged into 6 equal rows, each with 5 dots. So, we can put two rows in each of the 3 groups:

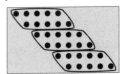

So, when the dots are divided into 3 equal groups, 5×2=**10** dots are in each group. 30÷3=10.

7. This pattern is made of 5 groups of 3 dots.

So, when the dots are divided into 5 equal groups, **3** dots are in each group. 15÷5=3.

8. The pattern is made of 6 rows, each with 12 dots. We can divide the dots into two equal groups as shown:

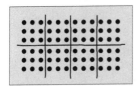

Each of the two equal groups can be split in half again to make 4 equal groups:

Finally, we split each of the four groups in half to create 8 equal groups:

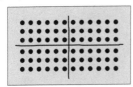

There are **9** dots in each group. 72÷8=9.

DIVISION

Missing Numbers

9. 8×$\boxed{3}$=24.

10. 9×$\boxed{4}$=36.

11. $\boxed{5}$×7=35.

12. 5×$\boxed{2}$=10.

13. $\boxed{9}$×3=27.

14. 2×$\boxed{10}$=20.

15. $\boxed{7}$×6=42.

16. 10×$\boxed{6}$=60.

17. 35÷7=$\boxed{5}$.

18. 60÷10=$\boxed{6}$.

19. 36÷9=$\boxed{4}$.

20. 42÷6=$\boxed{7}$.

21. 10÷5=$\boxed{2}$.

22. $27 \div 3 = \boxed{9}$.

23. $24 \div 8 = \boxed{3}$.

24. $20 \div 2 = \boxed{10}$.

25. The problems should be matched as follows:

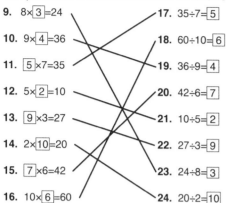

9. $8 \times \boxed{3} = 24$

10. $9 \times \boxed{4} = 36$

11. $\boxed{5} \times 7 = 35$

12. $5 \times \boxed{2} = 10$

13. $\boxed{9} \times 3 = 27$

14. $2 \times \boxed{10} = 20$

15. $\boxed{7} \times 6 = 42$

16. $10 \times \boxed{6} = 60$

17. $35 \div 7 = \boxed{5}$

18. $60 \div 10 = \boxed{6}$

19. $36 \div 9 = \boxed{4}$

20. $42 \div 6 = \boxed{7}$

21. $10 \div 5 = \boxed{2}$

22. $27 \div 3 = \boxed{9}$

23. $24 \div 8 = \boxed{3}$

24. $20 \div 2 = \boxed{10}$

DIVISION
Fact Wheels   40–41

26.   **27.**

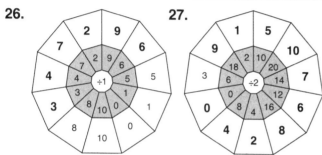

28.   **29.**

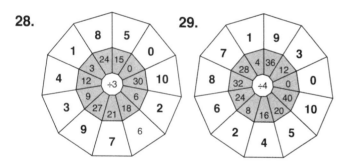

30.   **31.**

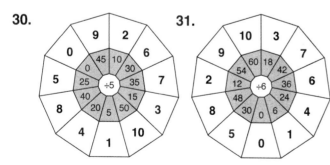

32.   **33.**

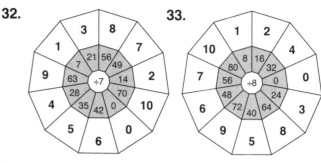

34.   **35.**

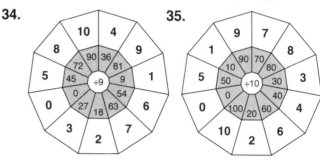

DIVISION
Basic Word Problems   42–43

36. Alex divides his 100 flowers into rows of 10 flowers each, so there are $100 \div 10 = \mathbf{10}$ rows.

37. Winnie divides her 32 flowers into 8 equal rows, so there are $32 \div 8 = \mathbf{4}$ flowers in each row.

38. Grogg divides his 56 flowers into 7 equal rows, so there are $56 \div 7 = \mathbf{8}$ flowers in each row.

39. When 40 cupcakes are divided equally among 4 little monsters, each little monster will get $40 \div 4 = \mathbf{10}$ cupcakes.

40. When 40 cupcakes are divided evenly among 5 little monsters, each little monster will get $40 \div 5 = \mathbf{8}$ cupcakes.

41. After Alex, Grogg, Lizzie, and Winnie each eat one cupcake, there are $40 - 4 = 36$ cupcakes left over. Then, the remaining 36 cupcakes are divided equally among 6 monsters. Each monster gets $36 \div 6 = 6$ cupcakes. So, Ralph gets **6** cupcakes.

42. Each bag of apples costs \$3, and $15 \div 3 = 5$. So, Lizzie can buy **5** bags of apples with \$15.

43. Each bag of bananas costs \$4, and $28 \div 4 = 7$. So, Fiona can buy **7** bags of bananas with \$28.

44. Each bag of mangoes costs \$5, and $30 \div 5 = 6$. So, Ms. Q. can buy **6** bags of mangoes with \$30.

45. Each bag of dewberries costs \$8. So, if Winnie can buy 5 bags of dewberries, she must have $8 \times 5 = 40$ dollars. Each bag of bananas costs \$4, and $40 \div 4 = 10$. So, Winnie can buy **10** bags of bananas.

— *or* —

For every \$8 bag of dewberries, Winnie can buy two \$4 bags of bananas. So, if Winnie can buy 5 bags of dewberries, she can buy $5 \times 2 = \mathbf{10}$ bags of bananas.

46. Alex could buy 24÷3=8 bags of apples with $24.
Alex could buy 24÷4=6 bags of bananas with $24.
So, Alex can buy 8−6=**2** more bags of apples than bags of bananas.

47. Grogg buys an equal number of bags of apples and bananas. So, for every bag of apples he buys, Grogg also buys a bag of bananas. Together, one bag of apples and one bag of bananas cost 3+4=7 dollars. 42÷7=6. So, Grogg could buy 6 bags of apples and 6 bags of bananas for $42. This is a total of 6+6=**12** bags of fruit.

DIVISION
Cross-Number Puzzles   44-46

48.

2	×	21	=	**42**
×	■	÷	■	−
15	÷	3	=	**5**
=	■	=	■	=
30	+	**7**	=	**37**

49.

56	−	50	=	**6**
÷	■	÷	■	+
7	+	5	=	**12**
=	■	=	■	=
8	+	**10**	=	**18**

50.

42	÷	6	=	**7**
−	■	+	■	+
36	÷	4	=	**9**
=	■	=	■	=
6	+	**10**	=	**16**

51.

56	−	54	=	**2**
÷	■	÷	■	+
8	+	6	=	**14**
=	■	=	■	=
7	+	**9**	=	**16**

52.

72	−	70	=	**2**
÷	■	÷	■	+
9	+	**7**	=	16
=	■	=	■	=
8	+	**10**	=	**18**

53.

11	×	7	=	**77**
×	■	×	■	+
6	÷	**2**	=	**3**
=	■	=	■	=
66	+	14	=	**80**

54.

36	÷	6	=	**6**
−	■	+	■	+
32	÷	**4**	=	**8**
=	■	=	■	=
4	+	10	=	**14**

55.

9	+	3	=	12
÷	■	−	■	÷
3	−	**1**	=	**2**
=	■	=	■	=
3	×	2	=	**6**

56.

32	÷	4	=	**8**
−	■	+	■	+
20	÷	2	=	**10**
=	■	=	■	=
12	+	6	=	**18**

57.

8	×	**5**	=	**40**
×	■	×	■	+
4	÷	2	=	**2**
=	■	=	■	=
32	+	10	=	**42**

58.

64	÷	**8**	=	**8**
−	■	+	■	+
60	÷	6	=	**10**
=	■	=	■	=
4	+	14	=	**18**

59.

6	×	6	=	36
÷	■	÷	■	÷
3	×	**3**	=	9
=	■	=	■	=
2	+	**2**	=	**4**

60.

45	÷	5	=	**9**
−	■	+	■	+
42	÷	**7**	=	6
=	■	=	■	=
3	+	**12**	=	**15**

61.

10	×	4	=	**40**
÷	■	÷	■	÷
5	×	**2**	=	**10**
=	■	=	■	=
2	+	**2**	=	**4**

62. We begin by evaluating the horizontal multiplication expression 3×40 and the vertical multiplication expression 3×35. Then, to fill in the center number, we look for a number that divides both 35 and 40. Since 35÷[5]=7 and 40÷[5]=8, we try 5.

3	×	40	=	**120**
×	■	÷	■	−
35	÷	**5**	=	
=	■	=	■	=
105	+		=	

Then, we can complete the puzzle as in the previous problems:

3	×	40	=	**120**
×	■	÷	■	−
35	÷	**5**	=	**7**
=	■	=	■	=
105	+	**8**	=	**113**

63. We begin by evaluating the horizontal multiplication expression 3×14 and finding the unknown number in the vertical multiplication equation 3×□=30.

3	×	14	=	**42**
×	■	÷	■	−
10	÷		=	
=	■	=	■	=
30	+		=	

To fill in the center number, we look for a number that divides both 10 and 14. Since 10÷$\boxed{2}$=5 and 14÷$\boxed{2}$=7, we try 2.

3	×	14	=	**42**
×	■	÷	■	−
10	÷	2	=	
=	■	=	■	=
30	+		=	

Then, we can complete the puzzle as in the previous problems:

3	×	14	=	**42**
×	■	÷	■	−
10	÷	2	=	5
=	■	=	■	=
30	+	7	=	37

64. To divide 500÷5, we find the number that can be multiplied by 5 to give us 500:

$\boxed{100}$×5=500.

So, 500÷5=**100**.

65. To divide 630÷7, we find the number that can be multiplied by 7 to give us 630. Since 9×7=63, $\boxed{90}$×7=630. So, 630÷7=**90**.

66. To divide 400÷8, we find the number that can be multiplied by 8 to give us 400. Since 5×8=40, $\boxed{50}$×8=400. So, 400÷8=**50**.

67. To divide 900÷3, we find the number that can be multiplied by 3 to give us 900. Since 3×3=9, $\boxed{300}$×3=900. So, 900÷3=**300**.

68. To divide 2,400÷4, we find the number that can be multiplied by 4 to give us 2,400. Since 6×4=24, $\boxed{600}$×4=2,400. So, 2,400÷4=**600**.

69. To divide 1,800÷9, we find the number that can be multiplied by 9 to give us 1,800. Since 2×9=18, $\boxed{200}$×9=1,800. So, 1,800÷9=**200**.

70. To divide 42,000÷6, we find the number that can be multiplied by 6 to give us 42,000. Since 7×6=42, $\boxed{7,000}$×6=42,000. So, 42,000÷6=**7,000**.

71. To divide 20,000÷5, we find the number that can be multiplied by 5 to give us 20,000. Since 4×5=20, $\boxed{4,000}$×5=20,000. So, 20,000÷5=**4,000**.

72. To divide 18,000÷20, we find the number that can be multiplied by 20 to give us 18,000. Since 9×2=18, $\boxed{900}$×20=18,000. So, 18,000÷20=**900**.

73. To divide 14,000÷700, we find the number that can be multiplied by 700 to give us 14,000. Since 2×7=14, $\boxed{20}$×700=14,000. So, 14,000÷700=**20**.

In each of the following long division problems, you may have used different steps to arrive at the same answer.

74. Step 1:
$$\begin{array}{r} 10 \\ 5{\overline{)}\,67} \\ -50 \\ \hline 17 \end{array}$$
Step 2:
$$\begin{array}{r} 10+3 \\ 5{\overline{)}\,67} \\ -50 \\ \hline 17 \\ -15 \\ \hline 2 \end{array}$$

So, the quotient of 67÷5 is 10+3=**13**. The remainder of 67÷5 is **2**.

75. Step 1:
$$\begin{array}{r} 10 \\ 3{\overline{)}\,43} \\ -30 \\ \hline 13 \end{array}$$
Step 2:
$$\begin{array}{r} 10+4 \\ 3{\overline{)}\,43} \\ -30 \\ \hline 13 \\ -12 \\ \hline 1 \end{array}$$

So, the quotient of 43÷3 is 10+4=**14**. The remainder of 43÷3 is **1**.

76.
$$\begin{array}{r} 9 \\ 7{\overline{)}\,64} \\ -63 \\ \hline 1 \end{array}$$

So, the quotient of 64÷7 is **9**. The remainder of 64÷7 is **1**.

77. Step 1:
$$\begin{array}{r} 10 \\ 4{\overline{)}\,76} \\ -40 \\ \hline 36 \end{array}$$
Step 2:
$$\begin{array}{r} 10+9 \\ 4{\overline{)}\,76} \\ -40 \\ \hline 36 \\ -36 \\ \hline 0 \end{array}$$

So, the quotient of 76÷4 is 10+9=**19**. The remainder of 76÷4 is **0**.

78. Step 1:
$$\begin{array}{r} 10 \\ 8{\overline{)}\,93} \\ -80 \\ \hline 13 \end{array}$$
Step 2:
$$\begin{array}{r} 10+1 \\ 8{\overline{)}\,93} \\ -80 \\ \hline 13 \\ -8 \\ \hline 5 \end{array}$$

So, the quotient of 93÷8 is 10+1=**11**. The remainder of 93÷8 is **5**.

79. Step 1:
$$
\begin{array}{r}
10 \\
7{\overline{\smash{\big)}\,98}} \\
-70 \\
\hline
28
\end{array}
$$

Step 2:
$$
\begin{array}{r}
10+4 \\
7{\overline{\smash{\big)}\,98}} \\
-70 \\
\hline
28 \\
-28 \\
\hline
0
\end{array}
$$

So, the quotient of 98÷7 is 10+4=**14**.
The remainder of 98÷7 is **0**.

80.
$$
\begin{array}{r}
1 \\
11{\overline{\smash{\big)}\,13}} \\
-11 \\
\hline
2
\end{array}
$$

So, the quotient of 13÷11 is **1**.
The remainder of 13÷11 is **2**.

81.
$$
\begin{array}{r}
1 \\
91{\overline{\smash{\big)}\,95}} \\
-91 \\
\hline
4
\end{array}
$$

So, the quotient of 95÷91 is **1**.
The remainder of 95÷91 is **4**.

82. Since 9 is greater than 5,
9 goes into 5 *zero* times.
$$
\begin{array}{r}
0 \\
9{\overline{\smash{\big)}\,5}} \\
-0 \\
\hline
5
\end{array}
$$

So, the quotient of 5÷9 is **0**.
The remainder of 5÷9 is **5**.

83.
$$
\begin{array}{r}
20 \\
6{\overline{\smash{\big)}\,125}} \\
-120 \\
\hline
5
\end{array}
$$

So, the quotient of 125÷6 is **20**.
The remainder of 125÷6 is **5**.

84. Step 1:
$$
\begin{array}{r}
5 \\
14{\overline{\smash{\big)}\,86}} \\
-70 \\
\hline
16
\end{array}
$$

Step 2:
$$
\begin{array}{r}
5+1 \\
14{\overline{\smash{\big)}\,86}} \\
-70 \\
\hline
16 \\
-14 \\
\hline
2
\end{array}
$$

So, the quotient of 86÷14 is 5+1=**6**.
The remainder of 86÷4 is **2**.

85. Step 1:
$$
\begin{array}{r}
300 \\
8{\overline{\smash{\big)}\,2507}} \\
-2400 \\
\hline
107
\end{array}
$$

Step 2:
$$
\begin{array}{r}
300+10 \\
8{\overline{\smash{\big)}\,2507}} \\
-2400 \\
\hline
107 \\
-80 \\
\hline
27
\end{array}
$$

Step 3:
$$
\begin{array}{r}
300+10+3 \\
8{\overline{\smash{\big)}\,2507}} \\
-2400 \\
\hline
107 \\
-80 \\
\hline
27 \\
-24 \\
\hline
3
\end{array}
$$

So, the quotient of 2,507÷8 is 300+10+3=**313**.
The remainder of 2,507÷8 is **3**.

86. The quotient of n÷8 is 15. This means that 8 goes into n 15 times. So, n is at least 15×8=120. The remainder of n÷8 is 6, so n is 6 more than 120. Since 120+6=126, n=**126**.

We can check our answer with long division:

Notice that n=(15×8)+6=126.

87. When 77 is divided by m, the quotient is 9 with remainder 5. This tells us that m goes into 77 nine times with 5 left over. So, m goes into 77−5=72 nine times with 0 left over. Since 9× 8 =72, m=**8**.

We can check our answer with long division:

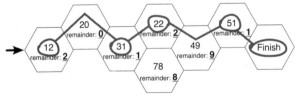

DIVISION Remainder Jump 50-53

Below, all the remainders are listed, and we show the only correct path for each maze.

88. Divisor: 10

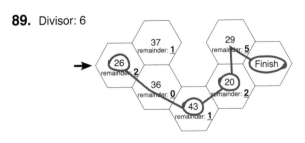

89. Divisor: 6

90. Divisor: 9

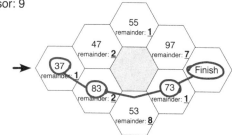

91. Divisor: 5

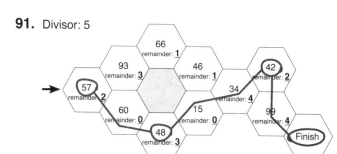

92. Divisor: 8

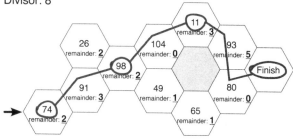

93. Divisor: 7

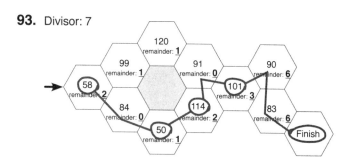

94. Divisor: 5

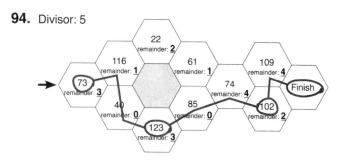

95. Divisor: 4

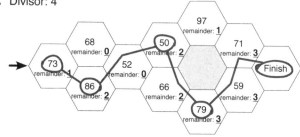

96. Divisor: 9

DIVISION
Perimeter and Area   57

97. A regular hexagon has 6 sides of equal length. So, to find the length of each side, we divide its perimeter by 6.

$$\begin{array}{r} 20+6 \\ 6 \overline{)156} \\ -120 \\ \hline 36 \\ -36 \\ \hline 0 \end{array}$$

156÷6 has remainder 0, so 156÷6=20+6=26.
The side length of a regular hexagon with perimeter 156 is **26**.

98. To find the area of a rectangle, we can multiply its height by its width. So, to find the height of a rectangle, we can divide its area by its width.

$$\begin{array}{r} 10+7 \\ 7 \overline{)119} \\ -70 \\ \hline 49 \\ -49 \\ \hline 0 \end{array}$$

119÷7 has remainder 0, so 119÷7=10+7=17.
The height of a rectangle with area 119 and width 7 is **17**.

99. A regular octagon has 8 sides of equal length. So, to find the length of each side, we divide its perimeter by 8.

$$\begin{array}{r} 20+3 \\ 8 \overline{)184} \\ -160 \\ \hline 24 \\ -24 \\ \hline 0 \end{array}$$

184÷8 has remainder 0, so 184÷8=20+3=23.
The length of each side of a regular hexagon with perimeter 184 is **23**.

100. To find the perimeter of a regular polygon, we can multiply the side length by the number of sides in the polygon. So, to find the number of sides in a regular polygon, we can divide its perimeter by its side length.

$$
\begin{array}{r}
10+2 \\
9{\overline{\smash{\big)}\,108}} \\
\underline{-90} \\
18 \\
\underline{-18} \\
0
\end{array}
$$

108÷9 has remainder 0, so 108÷9=10+2=12. Ralph's polygon has **12** sides.

A 12-sided polygon is called a *dodecagon*.

DIVISION
Clever Computations   58-59

101. To find the area of a rectangle, we can multiply its height by its width. So, to find the width of a rectangle, we can divide its area by its height.

$$
\begin{array}{r}
10+5 \\
7{\overline{\smash{\big)}\,105}} \\
\underline{-70} \\
35 \\
\underline{-35} \\
0
\end{array}
$$

105÷7 has remainder 0, so 105÷7=10+5=15.
The width of a rectangle with area 105 and height 7 is **15**.

102. To find the area of a rectangle, we can multiply its height by its width. So, to find the width of a rectangle, we can divide its area by its height.

$$
\begin{array}{r}
5+2 \\
15{\overline{\smash{\big)}\,105}} \\
\underline{-75} \\
30 \\
\underline{-30} \\
0
\end{array}
$$

105÷15 has remainder 0, so 105÷15=5+2=7.
The width of a rectangle with area 105 and height 15 is **7**.

— *or* —

To divide 105÷15, we find the number that can be multiplied by 15 to give us 105. In the previous problem, we saw that 105÷7=15, which means that 7×15=105. So, 7 can be multiplied by 15 to get 105. Therefore, 105÷15=7. The width of a rectangle with area 105 and height 15 is **7**.

103. To find the number of packs, we divide 117 markers into packs of 9.

$$
\begin{array}{r}
10+3 \\
9{\overline{\smash{\big)}\,117}} \\
\underline{-90} \\
27 \\
\underline{-27} \\
0
\end{array}
$$

117÷9 has remainder 0, so 117÷9=10+3=13.
Lizzie has **13** packs of markers.

104. To find the number of packs, we divide 117 markers into packs of 13. To divide 117÷13, we find the number that can be multiplied by 13 to get 117. In the previous problem, we saw that 117÷9=13, which means that 9×13=117. So, 9 can be multiplied by 13 to get 117. Therefore, 117÷13=9. Grogg has **9** packs of markers.

— *or* —

In the previous problem, we saw that 117 markers can be divided into 13 packs of 9. The number of markers in 13 packs of 9 markers is equal to the number of markers in **9** packs of 13 markers.

Notice that 117÷9=13 and 117÷13=9.

105. To find the number of piles, we divide 322 coins into piles of 23. To divide 322÷23, we find the number that can be multiplied by 23 to get 322. Captain Kraken can divide 322 coins into 23 piles of 14 coins, so 322÷14=23. This means that 14×23=322. So, 14 can be multiplied by 23 to get 322. Therefore, 322÷23=14. Captain Kraken can make **14** piles of 23 coins.

— *or* —

Captain Kraken can make 23 piles of 14 coins with 322 coins. The number of coins in 23 piles of 14 coins is equal to the number of coins in **14** piles of 23 coins.

Notice that 322÷14=23 and 322÷23=14.

106. To find the number of weeks in 168 days, we divide 168 days by the number of days in a week (7). To divide 168÷7, we find the number that can be multiplied by 7 to give us 168. We are given 7×24=168, so 24 is the number that can be multiplied by 7 to get 168. Therefore, 168÷7=24. There are **24** weeks in 168 days.

— *or* —

The number of *hours* in 7 days (24×7=168) is the same as the number of *days* in 24 weeks (7×24=168).

Notice that 168÷7=24 and 168÷24=7.

107. Since $n÷4=5$, we know that $n=5×4=20$.
So, $n÷5=20÷5=$**4**.

108. To divide $n÷239$, we find the number that can be multiplied by 239 to give us n. Since $n÷151=239$, we know that $n=239×151$. So, 151 is the number that can be multiplied by 239 to get n. Therefore, $n÷239=$**151**.

109. To divide $64÷n$, we find the number that can be multiplied by n to give us 64. Since $64÷m=n$, we know that $64=n×m$. So, m is the number that can be multiplied by n to get 64. Therefore, $64÷n=$**m**.

110. To divide $a÷c$, we find the number that can be multiplied by c to give us a. Since $a÷b=c$, we know that $a=c×b$. So, b is the number that can be multiplied by c to get a. Therefore, $a÷c=$**b**.

As long as a, b, and c are not zero, all these equations mean the same thing:

$$a=b×c,$$
$$a=c×b,$$
$$a÷b=c,$$
$$a÷c=b.$$

111. We first add 74+75 to find the total number of buttons Alex has, then divide by 7 to find the remainder. 74+75=149.

$$
\begin{array}{r}
20+1 \\
7\,)\overline{149} \\
-140 \\
\hline
9 \\
-7 \\
\hline
2
\end{array}
$$

The remainder is 2, so Alex will have **2** buttons left over.

— or —

Alex can first place the 74 green buttons into rows of 7. Since 74÷7 has remainder 4, Alex will have 4 green buttons left over.

Then, Alex can place the blue buttons into rows of 7. Since 75÷7 has remainder 5, Alex will have 5 blue buttons left over.

Alex has 4 green and 5 blue buttons left over, for a total of 9 extra buttons. Alex can make one more row of 7, and will have 9−7=**2** buttons left over.

Notice that (74+75)÷7 has the same remainder as (4+5)÷7.

112. We first add 36+37, then divide by 5 to find the remainder. 36+37=73.

$$
\begin{array}{r}
10+4 \\
5\,)\overline{73} \\
-50 \\
\hline
23 \\
-20 \\
\hline
3
\end{array}
$$

So, 73÷5 has remainder **3**.

— or —

We begin by finding the remainder when each number is divided by 5.
36÷5 has remainder 1.
37÷5 has remainder 2.
Then, we add the remainders.
(36+37)÷5 has the same remainder as (1+2)÷5.
1+2=3, and 3÷5 has remainder 3.
So, (36+37)÷5 has remainder **3**.

113. We first add 39+40, then divide by 6 to find the remainder. 39+40=79.

$$
\begin{array}{r}
10+3 \\
6\,)\overline{79} \\
-60 \\
\hline
19 \\
18 \\
\hline
1
\end{array}
$$

So, 79÷6 has remainder **1**.

— or —

We begin by finding the remainder when each number is divided by 6.
39÷6 has remainder 3.
40÷6 has remainder 4.

Then, we add the remainders.
(39+40)÷6 has the same remainder as (3+4)÷6.
3+4=7, and 7÷6 has remainder 1.
So, (39+40)÷6 has remainder **1**.

114. We first add 94+95+96, then divide by 9 to find the remainder. 94+95+96=285.

$$
\begin{array}{r}
30+1 \\
9\,)\overline{285} \\
-270 \\
\hline
15 \\
-9 \\
\hline
6
\end{array}
$$

So, 285÷9 has remainder **6**.

— or —

We begin by finding the remainder when each number is divided by 9.
94÷9 has remainder 4.
95÷9 has remainder 5.
96÷9 has remainder 6.
Then, we add the remainders.
(94+95+96)÷9 has the same remainder as (4+5+6)÷9.
4+5+6=15, and 15÷9 has remainder 6.
So, (94+95+96)÷9 has remainder **6**.

115. We begin by finding the remainder when each number is divided by 4.
46÷4 has remainder 2.
47÷4 has remainder 3.
48÷4 has remainder 0.
Then, we add the remainders.
(46+47+48)÷4 has the same remainder as (2+3+0)÷4.
2+3+0=5, and 5÷4 has remainder 1.
So, (46+47+48)÷4 has remainder **1**.

116. We begin by finding the remainder when each number is divided by 12.
13÷12 has remainder 1.
14÷12 has remainder 2.
15÷12 has remainder 3.
Then, we add the remainders.
(13+14+15)÷12 has the same remainder as (1+2+3)÷12.
1+2+3=6, and 6÷12 has remainder 6.
So, (13+14+15)÷12 has remainder **6**.

117. We begin by finding the remainder when each number is divided by 75.
187÷75 has remainder 37.
188÷75 has remainder 38.
189÷75 has remainder 39.
Then, we add the remainders.
(187+188+189)÷75 has the same remainder as (37+38+39)÷75.
Notice that 37+38=75, so 37+38+39 is 39 more than 75. This means (37+38+39)÷75 has remainder 39.
So, (187+188+189)÷75 has remainder **39**.

118. We first multiply 12×8 to find the total number of lemons, then divide by 7 to find the remainder. 12×8=96.

$$\begin{array}{r} 10+3 \\ 7\overline{)96} \\ -70 \\ \hline 26 \\ -21 \\ \hline 5 \end{array}$$

The remainder is 5, so there will be **5** lemons left over.

— *or* —

Each box contains 8 lemons. The bakers can use 7 lemons from each box to make a pie. There will be 1 lemon left in each of the 12 boxes, for a total of 12 extra lemons. The bakers can use 7 of these lemons to make 1 more pie. After making all possible pies, the bakers have 12−7=**5** lemons left over.

Notice that (12×8)÷7 has the same remainder as (5×1)÷7.

119. We first multiply 10×11, then divide by 8 to find the remainder. 10×11=110.

$$\begin{array}{r} 10+3 \\ 8\overline{)110} \\ -80 \\ \hline 30 \\ -24 \\ \hline 6 \end{array}$$

So, (10×11)÷8 has remainder **6**.

— *or* —

We begin by finding the remainder when each number is divided by 8.
10÷8 has remainder 2.
11÷8 has remainder 3.
Then, we multiply the remainders.
(10×11)÷8 has the same remainder as (2×3)÷8.
2×3=6, and 6÷8 has remainder 6.
So, (10×11)÷8 has remainder **6**.

120. We begin by finding the remainder when each number is divided by 6.
57÷6 has remainder 3.
58÷6 has remainder 4.
Then, we multiply the remainders.
(57×58)÷6 has the same remainder as (3×4)÷6.
3×4=12, and 12÷6 has remainder 0.
So, (57×58)÷6 has remainder **0**.

121. We begin by finding the remainder when each number is divided by 53.
54÷53 has remainder 1.
55÷53 has remainder 2.
Then, we multiply the remainders.
(54×55)÷53 has the same remainder as (1×2)÷53.
1×2=2, and 2÷53 has remainder 2.
So, (54×55)÷53 has remainder **2**.

122. We begin by finding the remainder when each number is divided by 5.
6÷5 has remainder 1.
7÷5 has remainder 2.
8÷5 has remainder 3.
Then, we multiply the remainders.
(6×7×8)÷5 has the same remainder as (1×2×3)÷5.
1×2×3=6, and 6÷5 has remainder 1.
So, (6×7×8)÷5 has remainder **1**.

123. First, we find the remainder when 11 is divided by 9.
11÷9 has a remainder of 2.
Then, we multiply the remainders.
(11×11×11)÷9 has the same remainder as (2×2×2)÷9.
2×2×2=8, and 8÷9 has remainder 8.
So, (11×11×11)÷9 has remainder **8**.

124. We begin by finding the remainder when each number is divided by 100.
103÷100 has remainder 3.
104÷100 has remainder 4.
105÷100 has remainder 5.
Then, we multiply the remainders.
(103×104×105)÷100 has the same remainder as (3×4×5)÷100.
3×4×5=60, and 60÷100 has remainder 60.
So, (103×104×105)÷100 has remainder **60**.

125. We first add all possible pairs of numbers, then find the remainder when each sum is divided by 5:
12+17=29 has remainder 4 when divided by 5.
12+23=35 has remainder 0 when divided by 5.
17+23=40 has remainder 0 when divided by 5.
Since 12+17=29 has remainder 4 when divided by 5, 12 and 17 cannot be in the same circle. We can place the numbers as shown:

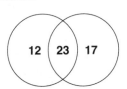

You may have switched the 12 and the 17.

— *or* —

We find the remainder when each number is divided by 5.
12÷5 has remainder 2.
17÷5 has remainder 2.
23÷5 has remainder 3.
Since (2+3)÷5 has remainder 0, we can place a number with remainder 2 and a number with remainder 3 in the same circle. Since (2+2)÷5 has remainder 4, we cannot place 12 and 17 in the same circle. So, 23 must be placed in the region where the circles overlap, with 12 and 17 placed on either side as shown above.

126. We begin by finding the remainder when each number is divided by 5.

11÷5 has remainder 1.
24÷5 has remainder 4.
39÷5 has remainder 4.

Since (1+4)÷5 has remainder 0, we can place a number with remainder 1 and a number with remainder 4 in the same circle. Since (4+4)÷5 has remainder 3, we cannot place two numbers with remainder 4 in the same circle. So, 11 must be placed in the region where the circles overlap, with 24 and 39 placed as shown below.

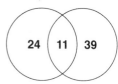

You may have switched the 24 and the 39.

We can check that the sum of numbers in the same circle have remainder 0 when divided by 5:
24+11=25 has remainder 0 when divided by 5. ✓
11+39=40 has remainder 0 when divided by 5. ✓

We use the strategies from the previous problems to find the following arrangements.

127. 13÷5 has remainder 3.
23÷5 has remainder 3.
27÷5 has remainder 2.

So, 27 must be placed in the region where the circles overlap, with 13 and 23 placed as shown below.

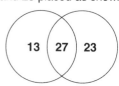

You may have switched the 13 and the 23.

128. 31÷5 has remainder 1.
36÷5 has remainder 1.
49÷5 has remainder 4.

So, 49 must be placed in the region where the circles overlap, with 31 and 36 placed as shown below.

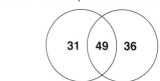

You may have switched the 31 and the 36.

129. 15÷8 has remainder 7.
23÷8 has remainder 7.
65÷8 has remainder 1.

So, 65 must be placed in the region where the circles overlap, with 15 and 23 placed as shown below.

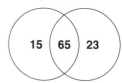

You may have switched the 15 and the 23.

130. 11÷8 has remainder 3.
21÷8 has remainder 5.
29÷8 has remainder 5.

So, 11 must be placed in the region where the circles overlap, with 21 and 29 placed as shown below.

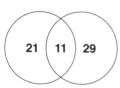

You may have switched the 21 and the 29.

131. 38÷8 has remainder 6.
46÷8 has remainder 6.
58÷8 has remainder 2.

So, 58 must be placed in the region where the circles overlap, with 38 and 46 placed as shown below.

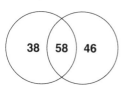

You may have switched the 38 and the 46.

132. 65÷8 has remainder 1.
73÷8 has remainder 1.
79÷8 has remainder 7.

So, 79 must be placed in the region where the circles overlap, with 65 and 73 placed as shown below.

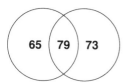

You may have switched the 65 and the 73.

*In the next four problems, we use the strategies from the previous problems, but we look to place the numbers so that each sum has remainder **3** when divided by 7.*

133. 64÷7 has remainder 1.
71÷7 has remainder 1.
79÷7 has remainder 2.

So, 79 must be placed in the region where the circles overlap, with 64 and 71 placed as shown below.

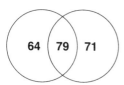

You may have switched the 64 and the 71.

134. 32÷7 has remainder 4.
48÷7 has remainder 6.
62÷7 has remainder 6.

Since (4+6)÷7 has remainder 3, we place 32 in the region where the circles overlap, with 48 and 62 placed as shown below.

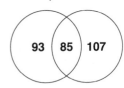

You may have switched the 48 and the 62.

135. 85÷7 has remainder 1.
93÷7 has remainder 2.
107÷7 has remainder 2.
So, 85 must be placed in the region where the circles overlap, with 93 and 107 placed as shown below.

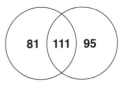

You may have switched the 93 and the 107.

136. 81÷7 has remainder 4.
95÷7 has remainder 4.
111÷7 has remainder 6.
Since (6+4)÷7 has remainder 3, we place 111 in the region where the circles overlap, with 81 and 95 placed as shown below.

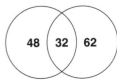

You may have switched the 81 and the 95.

DIVISION
Missing Numbers   64-65

137. 14+16=30, which has remainder 0 when divided by 10, so 16 cannot be in placed in a circle connected to 14. This leaves only one possible circle for the 16:

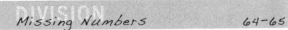

17+23=40, which has remainder 0 when divided by 10, so 17 and 23 cannot be in connected circles. This leaves only one possible circle for the 17:

Then, 15 can be placed in the remaining circle.

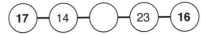

We can check that no two connected circles have a sum that has remainder 0 when divided by 10.

17+14=31 has remainder 1 when divided by 10. ✓
14+15=29 has remainder 9 when divided by 10. ✓
15+23=38 has remainder 8 when divided by 10. ✓
23+16=39 has remainder 9 when divided by 10. ✓

— *or* —

We find the remainder when each number is divided by 10. We start with the numbers in the diagram:
14÷10 has remainder 4.
23÷10 has remainder 3.
Then, we consider the missing numbers:
15÷10 has remainder 5.
16÷10 has remainder 6.
17÷10 has remainder 7.

We cannot place two numbers whose remainders have a sum of 10 in connected circles, because 10÷10 has remainder 0.
6+4=10, so 16 cannot be connected to 14.
7+3=10, so 17 cannot be connected to 23.
This leaves only one way to complete the diagram:

138. We find the remainder when each number is divided by 10. We start with the numbers in the diagram:
37÷10 has remainder 7.
19÷10 has remainder 9.
Then, we consider the missing numbers:
21÷10 has remainder 1.
22÷10 has remainder 2.
23÷10 has remainder 3.
Since 10÷10 has remainder 0, we cannot place two numbers whose remainders have a sum of 10 in connected circles.
1+9=10, so 21 cannot be connected to 19. This leaves only the top left circle for 21.
3+7=10, so 23 cannot be connected to 37. This leaves only the bottom right circle for 23.

The 22 fills the remaining circle as shown.

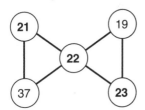

139. We find the remainder when each number is divided by 10. We start with the numbers in the diagram:

36÷10 has remainder 6.
48÷10 has remainder 8.
Then, we consider the missing numbers:
52÷10 has remainder 2.
53÷10 has remainder 3.
54÷10 has remainder 4.
Since 10÷10 has remainder 0, we cannot place two numbers whose remainders have a sum of 10 in connected circles.
2+8=10, so 52 cannot be connected to 48. This leaves only the top circle for 52.
4+6=10, so 54 cannot be connected to 36. This leaves only the top right circle for 54.
The 53 fills the remaining circle as shown.

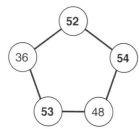

140. We find the remainder when each number is divided by 10. We start with the numbers in the diagram:

11÷10 has remainder 1.
18÷10 has remainder 8.
Then, we consider the missing numbers:
19÷10 has remainder 9.
20÷10 has remainder 0.
21÷10 has remainder 1.
Since 10÷10 has remainder 0, we cannot place two numbers whose remainders have a sum of 10 in connected circles.
1+9=10, so 19 cannot be connected to 11. This leaves only the far right circle for 19.
Similarly, 21 cannot be connected to 19. This leaves only the far left circle for 21.
The 20 fills the remaining circle as shown.

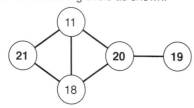

141. We find the remainder when each number is divided by 5. We start with the numbers in the diagram:

12÷5 has remainder 2.
29÷5 has remainder 4.
Then, we consider the missing numbers:
26÷5 has remainder 1.
27÷5 has remainder 2.
28÷5 has remainder 3.
Since 5÷5 has remainder 0, we cannot place two numbers whose remainders have a sum of 5 in connected circles.

1+4=5, so 26 cannot be connected to 29. This leaves only the far left circle for 26.
3+2=5, so 28 cannot be connected to 12. This leaves only the far right circle for 28.
The 27 fills the remaining circle as shown.

142. We find the remainder when each number is divided by 5. We start with the numbers in the diagram:

42÷5 has remainder 2.
36÷5 has remainder 1.
Then, we consider the missing numbers:
32÷5 has remainder 2.
33÷5 has remainder 3.
34÷5 has remainder 4.
Since 5÷5 has remainder 0, we cannot place two numbers whose remainders have a sum of 5 in connected circles.
2+3=5, so 33 cannot be connected to 42. This leaves only the far right circle for 33.
1+4=5, so 34 cannot be connected to 36. This leaves only the far left circle for 34.
The 32 fills the remaining circle as shown.

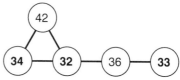

143. We find the remainder when each number is divided by 5. We start with the numbers in the diagram:

23÷5 has remainder 3.
35÷5 has remainder 0.
Then, we consider the missing numbers:
15÷5 has remainder 0.
16÷5 has remainder 1.
17÷5 has remainder 2.
Since 5÷5 has remainder 0, we cannot place two numbers whose remainders have a sum of 5 in connected circles. 2+3=5, so 17 cannot be connected to 23. This leaves only the top right circle for 17.
Since 0÷5 has remainder 0, we cannot place two numbers whose remainders have a sum of 0 in connected circles. So, 15 cannot be connected to 35. This leaves only the bottom right circle for 15.
The 16 fills the remaining circle as shown.

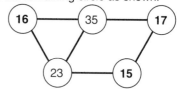

144. We find the remainder when each number is divided by 5. We start with the numbers in the diagram:

16÷5 has remainder 1.
38÷5 has remainder 3.
Then, we consider the missing numbers:
41÷5 has remainder 1.
42÷5 has remainder 2.

43÷5 has remainder 3.

Since 5÷5 has remainder 0, we cannot place two numbers whose remainders have a sum of 5 in connected circles.

2+3=5, so 42 cannot be connected to 38. So, 42 could only be in the bottom center circle or the far right circle. Similarly, 43 cannot be connected to 42. The only two blank circles not connected are the far left and far right. 42 cannot be in the far left circle, so we place 42 in the far right circle and 43 in the far left circle. The 41 fills the remaining circle as shown.

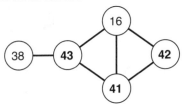

145. We find the remainder when each number is divided by 6. We start with the numbers in the diagram:

22÷6 has remainder 4.

27÷6 has remainder 3.

Then, we consider the missing numbers:

32÷6 has remainder 2.

33÷6 has remainder 3.

34÷6 has remainder 4.

Since 6÷6 has remainder 0, we cannot place two numbers whose remainders have a sum of 6 in connected circles.

2+4=6, so 32 cannot be connected to 22. So, 32 can only be placed in one of the bottom two circles. Similarly, 34 cannot be connected to 32. The only two blank circles not connected are the top left and bottom right. Since 32 cannot be placed in the top left circle, we place 32 in the bottom right circle and 34 in the top left circle. The 33 fills the remaining circle as shown.

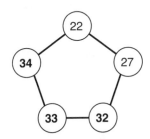

146. We find the remainder when each number is divided by 6. We start with the numbers in the diagram:

55÷6 has remainder 1.

68÷6 has remainder 2.

Then, we consider the missing numbers:

71÷6 has remainder 5.

72÷6 has remainder 0.

73÷6 has remainder 1.

Since 6÷6 has remainder 0, we cannot place two numbers whose remainders have a sum of 6 in connected circles.

1+5=6, so 71 cannot be connected to 55. So, 71 can only be placed in the center circle or the bottom right circle. Similarly, 73 cannot be connected to 71. The only

two blank circles not connected are the bottom left and bottom right.

71 cannot be in the bottom left circle, so we place 71 in the bottom right circle and 73 in the bottom left circle. The 72 fills the remaining circle as shown.

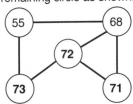

147. We use long division to divide 234÷9:

$$9\overline{)234} = \begin{array}{r} 20+6 \\ \hline 234 \\ -180 \\ \hline 54 \\ -54 \\ \hline 0 \end{array}$$

234÷9 has remainder 0, so 234÷9=20+6=26. So, there will be **26** students in each classroom.

148. Lizzie has 7×9=63 stickers. If she divides her stickers equally onto 9 pages, there will be 63÷9=**7** stickers on each page.

— *or* —

7×9=9×7, so 7 pages with 9 stickers contain the same number of stickers as 9 pages with **7** stickers.

149. We find the perimeter of the nonagon, then divide by 3 to find the side length of the triangle. The nonagon has perimeter 9×13=117. We can use long division to divide 117÷3:

$$3\overline{)117} = \begin{array}{r} 30+9 \\ \hline 117 \\ -90 \\ \hline 27 \\ -27 \\ \hline 0 \end{array}$$

107÷3=39, so the triangle has sides of length **39**.

— *or* —

The nonagon has 9 sides, and the triangle has 3 sides. So, the nonagon has 3 sides for every side of the triangle. We can think of the sides of the regular nonagon "folding" in to transform it into an equilateral triangle.

Since the perimeters of the triangle and the nonagon are equal, the side length of the triangle equals the sum of the lengths of three sides of the nonagon. So, the triangle has side length 13+13+13=**39**.

150. Grogg can split his 78 crayons into groups of 3, with each group including one red and two purple crayons. There are 78÷3 groups.

$$
\begin{array}{r}
20+6 \\
3{\overline{\smash{\big)}\,78}} \\
\underline{-60} \\
18 \\
\underline{-18} \\
0
\end{array}
$$

So, there are 20+6=26 groups of 3 crayons. Each group has two purple crayons, so Grogg has 26×2=**52** purple crayons.

151. We can multiply 7×36 to find the total number of rubies, then divide by 4 to figure out how many rubies will be in each chest. 7×36=252. We use long division to divide 252÷4:

$$
\begin{array}{r}
60+3 \\
4{\overline{\smash{\big)}\,252}} \\
\underline{-240} \\
12 \\
\underline{-12} \\
0
\end{array}
$$

So, there are 60+3=**63** rubies in each chest.

— *or* —

Captain Kraken can take each bag of 36 rubies and place 36÷4=9 rubies in each chest. There are 7 bags of rubies, so there will be 9×7=**63** rubies in each chest.

152. We can multiply 6×20, then divide by 17 to find the remainder. 6×20=120.

$$
\begin{array}{r}
5+2 \\
17{\overline{\smash{\big)}\,120}} \\
\underline{-85} \\
35 \\
\underline{-34} \\
1
\end{array}
$$

So, Ms. Q. will have **1** pencil left over.

— *or* —

Each pack has 20 pencils. Ms. Q. can give one pencil from every pack to each of 17 students. She will have 3 pencils left in each pack. There are 6 packs, so there will be 6×3=18 pencils left over. Ms. Q. can give one more pencil to each of 17 students and have 18−17=**1** pencil left over. Notice that (6×20)÷17 has the same remainder as (6×3)÷17.

153. We can find the area of the rectangle, then divide by 9. The area of the rectangle is 9×16=144 square units. We use long division to divide 144÷9:

$$
\begin{array}{r}
10+6 \\
9{\overline{\smash{\big)}\,144}} \\
\underline{-90} \\
54 \\
\underline{-54} \\
0
\end{array}
$$

So, each small rectangle has an area of 10+6=**16** square units.

— *or* —

The area of the large rectangle is 9×16. To find the area of a small rectangle, we divide (9×16)÷9. So, we find the number that can be multiplied by 9 to get 9×16. That number is 16. So, each small rectangle has an area of **16** square units.

154. We can multiply 6×17, then divide by 3 to find the number of gumballs in each pile. 6×17=102. We use long division to divide 102÷3:

$$
\begin{array}{r}
30+4 \\
3{\overline{\smash{\big)}\,102}} \\
\underline{-90} \\
12 \\
\underline{-12} \\
0
\end{array}
$$

So, there are 30+4=**34** gumballs in each pile.

— *or* —

Grogg has 6 piles of 17 gumballs. He can combine pairs of the six original piles to make 3 new piles of 17+17 gumballs. So, there are 17+17=**34** gumballs in each pile.

155. We can divide 78÷14 and find the remainder:

$$
\begin{array}{r}
5 \\
14{\overline{\smash{\big)}\,78}} \\
\underline{-70} \\
8
\end{array}
$$

So, there are 5 complete stacks, with **8** blocks left over

— *or* —

78 blocks make 11 complete stacks of 7, with one block left over. We can combine 10 of these stacks of 7 to make 5 stacks of 14 blocks. This leaves one stack of 7, plus one block left over, for a total of 7+1=**8** blocks left over.

156. We can find the number of students, then divide the number of toothpicks by the total number of students. There are 7 teams, with 5 students on each team, for a total of 7×5=35 students.
We use long division to divide 210÷35:

$$
\begin{array}{r}
5+1 \\
35{\overline{\smash{\big)}\,210}} \\
\underline{-175} \\
35 \\
\underline{-35} \\
0
\end{array}
$$

So, there are 5+1=**6** toothpicks for each student.

— *or* —

We can divide the toothpicks by team, then by student. There are 7 teams, so each team gets 210÷7 toothpicks. 210÷7=30, so each team gets 30 toothpicks. There are 5 students on each team, so each student gets 30÷5=**6** toothpicks.

157. We can find the total number of popsicle sticks, then divide by 7 to find the remainder. 50+58+66+74=248. We use long division to divide 248÷7:

$$
\begin{array}{r}
30{+}5 \\
7\overline{)\,248} \\
-210 \\
\hline
38 \\
-35 \\
\hline
3
\end{array}
$$

The remainder is 3, so **3** popsicle sticks are left over.

— or —

First, we find the remainder when the number of each color is divided by 7:
(Blue) 50÷7 has remainder 1.
(Purple) 58÷7 has remainder 2.
(Green) 66÷7 has remainder 3.
(Pink) 74÷7 has remainder 4.
Then, we add the remainders. (50+58+66+74)÷7 has the same remainder as (1+2+3+4)÷7.
1+2+3+4=10, and 10÷7 has remainder 3. So, there will be **3** popsicle sticks left over.

158. To find the number of cookies Alex can decorate with 100 chocolate chips, we divide 100÷3.

$$
\begin{array}{r}
30{+}3 \\
3\overline{)\,100} \\
-90 \\
\hline
10 \\
-9 \\
\hline
1
\end{array}
$$

So, with 100 chocolate chips, Alex can decorate 30+3=33 cookies, with 1 chocolate chip left over.
To find the number of cookies Alex can decorate with 125 cinnamon candies, we divide 125÷4:

$$
\begin{array}{r}
30{+}1 \\
4\overline{)\,125} \\
-120 \\
\hline
5 \\
-4 \\
\hline
1
\end{array}
$$

So, with 125 cinnamon candies, Alex can decorate 30+1=31 cookies, with 1 cinnamon candy left over.
Alex has enough chocolate chips to decorate 33 cookies, but only enough cinnamon candies to decorate 31 cookies. So, Alex can decorate **31** cookies.

159. To find the number of buses, we divide the 223 little monsters into groups of 40. We use long division to divide 223÷40:

$$
\begin{array}{r}
5 \\
40\overline{)\,223} \\
-200 \\
\hline
23
\end{array}
$$

So, the little monsters can be divided into 5 buses full of 40 little monsters, with 23 little monsters left over. To take those 23 left over little monsters on the field trip, one more bus is needed! All together, 5+1=**6** buses are needed to take 223 little monsters on a field trip.

160. First, we divide 41÷7:

$$
\begin{array}{r}
5 \\
7\overline{)\,41} \\
-35 \\
\hline
6
\end{array}
$$

41÷7 has quotient 5 and remainder 6. So, $a=5$ and $b=6$. We are looking for a number that has quotient 6 and remainder 5 when divided by 7. This means that 7 goes into our number 6 times, with 5 left over. 7×6 is 42, plus 5 extra is 47. So, **47** has quotient 6 and remainder 5 when divided by 7.

161. There are 7 days in a week, so a leap year has 366÷7 weeks. We use long division to divide 366÷7:

$$
\begin{array}{r}
50{+}2 \\
7\overline{)\,366} \\
-350 \\
\hline
16 \\
-14 \\
\hline
2
\end{array}
$$

So, a leap year has 52 weeks, plus 2 extra days. Kara was born on a Sunday, so we can divide the year she was born into 52 weeks that begin on Sunday and end on a Saturday. The last two days of the year are Sunday and Monday, so her first birthday will be on **Tuesday**, January 1st.

162. Grogg's name has three G's. We can find the number of times Grogg wrote his name by dividing the number of G's he wrote by 3. We use long division to divide 162÷3:

$$
\begin{array}{r}
50{+}4 \\
3\overline{)\,162} \\
-150 \\
\hline
12 \\
-12 \\
\hline
0
\end{array}
$$

So, Grogg wrote his name 50+4=54 times. Grogg's name has one O, so there are **54** O's on Grogg's paper.

MEASUREMENT

Introduction   page 71

There are many good answers to Problems 1-5. The answers below may be different from your answers, depending on how you thought about each item.

1. One useful measure of a kite string is its **length**. You might also be interested in knowing the **price** of a kite string if you are buying one.

2. The **price** of a refrigerator is important if you are buying one. The **length** of a refrigerator will help you determine if it will fit in your kitchen. The **temperature** inside a refrigerator is important for keeping food fresh. The **volume** of a refrigerator lets you know how much space it takes up. The capacity of a refrigerator is the volume inside the refrigerator, which helps you know how much it will hold. If you need to move or ship a refrigerator, it may be useful to know its **weight**.

3. The **length** of a footrace describes how far runners have to go. It is useful to know how much **time** it takes each runner to finish the race. It is also useful to know the **temperature** if you are in the race and deciding what to wear. Some footraces have an entrance fee, which is the **price** you must pay to enter the race.

4. The **temperature** of a swimming pool describes how warm or cold the water is. **Length** can be used to describe how far it is from one end to the other, how wide the pool is, or even how deep the water is. If you are buying a pool, the **price** is important. The **volume** of water contained in a pool is a useful measure if you are adding chemicals like chlorine. You could measure a pool's weight, but it is not a particularly useful measure of a pool.

5. You may have connected movie to length. However, the *length* of a movie usually means how much **time** the movie lasts from start to finish, not the kind of length you measure with a ruler. The **price** of a movie is also useful to know if you are paying to watch it.

For problems 1-5, we connected the items below to the measures on the right. Your answers may be different.

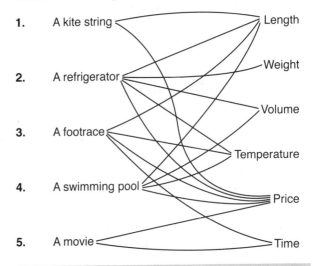

MEASUREMENT

Units of Length & Distance   72-73

6. There are 12 inches in 1 foot. So, 7 feet equals 7×12=84 inches. This means 7 feet 3 inches equals 84+3=87 inches. Barry is **87** inches tall.

7. There are 100 centimeters in one meter. So, three meters equals 3×100=300 centimeters. Half of 100 is 50. So, half of a meter is 50 centimeters. Three and a half meters equals 300+50=**350** centimeters.

8. We are looking for the length of each piece in inches. So, we convert the length of the rope into inches. There are 12 inches in a foot, so 5 feet equals 5×12=60 inches. Lizzie cuts the rope into 3 equal pieces, so each piece is 60÷3=**20** inches long.

9. We are looking for the side length of the square in centimeters. So, we convert the perimeter of the square into centimeters. One meter equals 100 centimeters. A square has four equal sides. So, each side of the square is 100÷4=**25** centimeters long.

10. We are looking for the length of one lap in feet. Since one mile equals 5,280 feet, 3 laps equals 5,280 feet. So, one lap around the track is 5,280÷3=**1,760** feet long.

11. Seven blocks can be stacked to a height of 40 cm. We want to make a stack that is 2 meters tall. There are 2×100=200 centimeters in two meters. Since 200÷40=5, it takes five 40-centimeter stacks of blocks to reach a height of 200 centimeters.

    Each 40-centimeter stack requires 7 blocks. So, the 200 cm stack requires 5 stacks of 7 blocks. This means a total of 5×7=**35** blocks are needed make a stack that is 2 meters tall.

12. We will use a diagram and a variable to solve this problem. If we label the height of the rectangle below h, we can label its width $h+h$, because the problem states that the width of the rectangle is double its height.

$h+h$

h

The perimeter of the rectangle above is
$h+(h+h)+h+(h+h)=h+h+h+h+h+h=6\times h$.
We know that the perimeter is 54 inches, so $6\times h=54$.
Since $6\times \boxed{9}=54$, we know $h=9$. So, the height of the rectangle is **9 inches**.

13. The first cup is 6 inches tall. A 2-cup stack is 8 inches tall, so each extra cup adds $8-6=2$ inches to the height of the cup tower. Three feet equals $3\times12=36$ inches. To make a tower that is 36 inches tall, we need to add $36-6=30$ inches after the first cup. Since each cup adds 2 inches, it will take $30\div2=15$ cups to add 30 inches. Including the original cup, it takes $15+1=$**16** cups to make a tower that is 3 feet tall.

14. One mile is 5,280 feet, which equals 1,760 yards. So, 1 mile is longer than 30 yards, 100 feet, or 1,000 inches (since 1 foot is more than 1 inch, we know that 5,280 feet is longer than 1,000 inches). This makes 1 mile the longest of the four distances.

To compare 30 yards, 100 feet, and 1,000 inches, we convert each distance to the same unit. We will convert all three measurements to inches.

1,000 inches is already in inches.

Since 1 foot equals 12 inches, 100 feet equals $100\times12=1,200$ inches.

Since 1 yard equals 3 feet, 30 yards equals $30\times3=90$ feet. Then, since 1 foot equals 12 inches, 90 feet equals $90\times12=1,080$ inches.

Since 1,200 inches is more than 1,080 inches, which is more than 1,000 inches, we know that 100 feet is more than 30 yards, which is more than 1,000 inches.

In order from longest to shortest, we have **1 mi, 100 ft, 30 yd, 1,000 in**.

15. The distance between B and C is **3 inches**.

16. Points **B** and **D** are 4 inches apart.

17. Points **A** and **C** are 5 inches apart.

18. Points **A** and **D** are 6 inches apart.

19. It is **11 centimeters** from Fran's house to Jo's house.

20. From Ed's to Gil's is 7 cm, from Gil's to Fran's is 2 cm, from Fran's to Hank's is 8 cm, from Hank's to Kiki's is 4 cm, and from Kiki's to Jo's is 1 cm. The total distance Bobby travels is $7+2+8+4+1=$**22 centimeters**.

21.
1 cm: **J & K**	2 cm: **F & G**	3 cm: **H & J**
4 cm: **H & K**	5 cm: **E & F**	6 cm: **G & H**
7 cm: **E & G**	8 cm: **F & H**	9 cm: **G & J**

22. Houses G and K are 10 cm apart, F and J are 11 cm apart, F and K are 12 cm apart, and E and H are 13 cm apart. However, no two houses are separated by **14 cm**.

23. We call each walk Bobby takes from one house to the other a "trip". Beginning at one house, Bobby needs to make 5 trips to visit the other 5 houses. We want each of these trips to be as long as possible to reach 62 cm. We will assume that one of Bobby's trips is from Ed's house to Kiki's house, since that is the longest possible trip.

The farthest house from Ed's (besides Kiki's) is Jo's. The farthest house from Kiki's (besides Ed's) is Fran's. So, to make the distance as long as possible, Bobby's walk will include the trip between E and J, and the trip between K and F as shown below. Bobby will always walk a straight line. The arrows are curved to make the diagram clear.

This leaves only two houses for Bobby to visit: Gil's and Hank's. Gil's house is farther from Jo's, and Hank's house is farther from Fran's. So, Bobby's walk will include the trip between J and G, and the trip between F and H as shown below.

Bobby can make the walk above starting at Gil's and ending at Hank's (**G→J→E→K→F→H**), or starting at Hank's and ending at Gil's (**H→F→K→E→J→G**) for a total distance of $8+12+17+16+9=62$ cm.

There are 8 ways Bobby can travel a total distance of 62 centimeters. Four begin at Gil's and end at Hank's. The other four are the same walk reversed:

G→J→E→K→F→H	H→F→K→E→J→G
G→J→F→K→E→H	H→E→K→F→J→G
G→K→E→J→F→H	H→F→J→E→K→G
G→K→F→J→E→H	H→E→J→F→K→G

There is no order that requires Bobby to travel more than 62 cm to visit all six friends once.

24. The side length of the regular pentagon is 2 cm.

2 cm

So, its perimeter is $5\times2=$**10 cm**.

25. The lengths of the sides of the rectangle are labeled below.

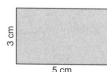

3 cm

5 cm

So, its perimeter is $3+5+3+5=$**16 cm**.

26. The lengths of the sides of the right triangle are labeled below.

So, its perimeter is 3+4+5=**12 cm**.

27. The lengths of the sides of the pentagon are labeled below.

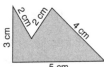

So, its perimeter is 3+2+2+4+5=**16 cm**.

28. The lengths of the sides of the quadrilateral ABCD are labeled below.

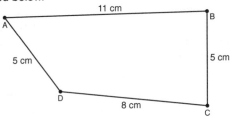

So, its perimeter is 5+11+5+8=**29 cm**.

29. The lengths of the sides of the triangle ACD are labeled below.

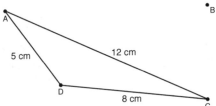

So, its perimeter is 12+8+5=**25 cm**.

30. There are four different triangles that can be made by connecting three of the points in the diagram: ABC, ABD, ACD, and BCD. Triangle ABC has the greatest perimeter.

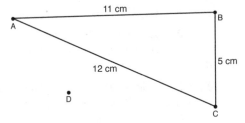

Triangle ABC has perimeter 11+5+12=**28 cm**.

31. The lengths of the sides of the quadrilateral EFGH are labeled below.

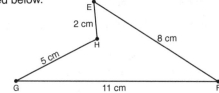

So, its perimeter is 5+2+8+11=**26 cm**.

32. There are four different triangles that can be made by connecting three of the points in the diagram: EFG, EFH, EGH, and FGH. Triangle EGH has a perimeter of 13 cm.

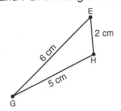

So, points **E**, **G**, and **H** can be connected to make a triangle with a perimeter of 13 cm.

33. Of the four triangles that can be made by connecting three of the points in the diagram, triangle EFG has the greatest perimeter.

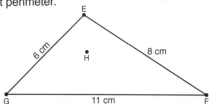

Triangle EFG has perimeter 6+8+11=**25 cm**.

MEASUREMENT

Project: Pendulum 78

Pulling the pendulum farther back does not significantly affect its period for small or medium swings. When you pull the pendulum back more, it must swing farther, but it also travels faster, so the period stays about the same. For large swings, the period increases. For example, if you measure the period of a pendulum that is pulled to the height of the doorway, its period will be longer than a pendulum of the same length that swings gently.

Changing the length of the pendulum changes its period. The longer the string, the longer the pendulum's period. However, doubling the length of a pendulum will not double its period. To double the period of a pendulum, you need to make the pendulum 4 times as long!

For small swings, the length of a pendulum with period 2 seconds is almost exactly 100 centimeters (1 meter). Here are some more lengths and periods to check against your results:

Length (cm)	Period (seconds)
160	2.54
140	2.37
120	2.20
100	2.01
80	1.80
60	1.55
40	1.27

For a pendulum pulled to the height of the doorway, an 80-centimeter pendulum will have a period of about 2 seconds. Your results may not match these exactly, but they should be pretty close.

Looking to learn more about pendulums? Explore whether changing the weight affects the period of your pendulum.

34. There are 1,000 grams in 1 kilogram, so there are 4×1,000=**4,000** grams in four kilograms.

35. Since there are 16 ounces in 1 pound, three pounds equals 3×16=48 ounces. So, 3 pounds 9 ounces equals 48+9=**57** ounces.

36. The combined weight of a 50-ounce steak and a 14-ounce baked potato is 50+14=64 ounces. Since there are 16 ounces in 1 pound, we divide 64 by 16 to convert 64 ounces to pounds. 64÷16=4, so there are 4 pounds in 64 ounces. The whole meal weighs **4** pounds.

37. We can find the weight of Lizzie's textbooks in ounces, add the weights, then convert back to pounds. Since 1 pound equals 16 ounces, each of Lizzie's textbooks weighs 16+5=21 ounces. She has five textbooks, for a total weight of 5×21=105 ounces. Since there are 16 ounces in 1 pound, we divide 105 by 16 to convert 105 ounces to pounds. Since 105÷16 has quotient 6 and remainder 9, there are 6 whole pounds in 105 ounces, with 9 ounces left over. We can write this as a mixed measure in pounds and ounces: **6 lb 9 oz**.

— *or* —

We can find the pounds and ounces separately. Each of Lizzie's books weighs 1 lb 5 oz. Lizzie has five books, so the total weight is 5×1=5 pounds, plus 5×5=25 ounces. So, Lizzie's books weigh 5 lb + 25 oz. When we write a mixed measure, the number of ounces must be less than 16. Since there are 16 ounces in a pound, 25 oz is the same as 1 lb+9 oz.
So, 5 lb+25 oz = 5 lb+1 lb+9 oz = 6 lb+9 oz, or **6 lb 9 oz**.

38. The arrow is 4 tick marks past the 3. So, the weight of the sack is **3 lb 4 oz**.

39. The octapug weighs the same as a 35-gram weight plus a 20-gram weight. So, the weight of the octapug is 35+20=**55 grams**.

40. The pandakeet weighs the same as an 18-gram weight plus a 19-gram weight. So, the weight of the pandakeet is 18+19=**37 grams**.

41. The arrow is 6 tick marks past the 1. So, the weight of the sack is **1 lb 6 oz**.

42. On the first scale, the arrow is at the 12th tick mark. So, the weight of the sack on the first scale is 12 oz. On the second scale, the arrow is at the 5th tick mark past 1. So, the weight on the second scale is 1 lb 5 oz. Since there are 16 ounces in 1 pound, we can convert 1 lb 5 oz to 16+5=21 oz. Adding the weights, we get 12+21=33 ounces. There are 16 ounces in 1 pound, so we divide 33 by 16 to convert 33 ounces to pounds. Since 33÷16 has quotient 2 and remainder 1, there are 2 whole pounds in 33 ounces, with 1 ounce left over. We can write this as a mixed measure in pounds and ounces: **2 lb 1 oz**.

— *or* —

We can add the ounces without converting. 1 lb+5 oz+12 oz = 1 lb+17 oz. When we write a mixed measure, the number of ounces must be less than 16. Since there are 16 ounces in a pound, 17 oz is the same as 1 lb+1 oz. So, 1 lb+17 oz=1 lb+1 lb+1 oz =2 lb+1 oz, or **2 lb 1 oz**.

43. For the scale to balance, the weight on the left side must equal the weight on the right side. So, half of the total weight must be placed on each side. The octapug weighs 55 grams, and the pandakeet weighs 37 grams. All together, the five items weigh 23+15+10+55+37=140 grams. Half of 140 is 70, so we must place 70 grams on each side of the scale. The octapug weighs 55 grams, so it can be placed with the 15-gram weight to equal 70 grams. The 37-gram pandakeet can be placed with the 23-gram weight and the 10-gram weight to equal 70 grams. The scale below shows how all five items could be placed. You may have switched the sides.

44. The top scale displays the weight of the sack only. The sack weighs 1 lb 6 oz. The bottom scale displays the weight of the top scale plus the sack. Together, the scale and the sack weigh 3 lb 11 oz. To find the weight of the scale by itself, we can subtract the weight of the sack from the combined weight of the sack and the scale. To subtract 1 lb 6 oz from 3 lb 11 oz, it is easiest to subtract the pounds and ounces separately.

$$
\begin{array}{r r}
3\text{ lb} & 11\text{ oz} \\
-\ 1\text{ lb} & 6\text{ oz} \\
\hline
2\text{ lb} & 5\text{ oz}
\end{array}
$$

So, one scale weighs **2 lb 5 oz**.

45. There are 2 cups in a pint, and each cup equals 8 fluid ounces. So, a pint equals 2×8=16 fluid ounces. 1 pt=**16** fl oz.

There are 2 pints in a quart, and we just computed that 1 pint equals 16 fluid ounces. So, a quart equals 2×16=32 fluid ounces. 1 qt=**32** fl oz.

There are 4 quarts in a gallon, and we just computed that 1 quart equals 32 fluid ounces. So, a gallon equals 4×32=128 fl oz. 1 gal=**128** fl oz.

46. There are 1,000 milliliters in 1 liter. So, 2 liters equals 2×1,000=2,000 milliliters. Half of 1,000 is 500. So, a half-liter equals 500 milliliters. Therefore, two and a half liters equals 2,000+500=**2,500 mL**.

47. We learned in Problem 45 that a gallon equals 128 fluid ounces. Half of 128 is 64. So, a half-gallon equals 64 fluid ounces. A pint equals 16 fluid ounces. Half of 16 is 8. So, a half-pint equals 8 fluid ounces. This means a half-gallon plus a half-pint is 64+8=**72 fl oz**.

48. If each glass holds 10 fluid ounces, then 8 glasses will hold 8×10=80 fluid ounces. In Problem 45, we found that a pint equals 16 fluid ounces. So, to figure out how many pints it takes to equal 80 fluid ounces, we divide 80 by 16. Since 80÷16=5, it will take **5 pints** to fill all 8 glasses.

49. The water reaches the second tick mark past 40, so the graduated cylinder holds **42 mL**.

50. The water reaches one tick mark past 12, so the measuring cup holds 13 fl oz. Since 1 cup equals 8 fluid ounces, 13 fluid ounces is 5 fluid ounces more than 1 cup. So, the cup holds **1 cup 5 fl oz**.

51. We can add the fluid ounces in each measuring cup to find the total, then find the remainder when the total is divided by the number of fluid ounces in each jar (5). The measuring cups hold 16 fl oz, 9 fl oz, and 11 fl oz. All together, this is 16+9+11=36 fluid ounces. Since 36÷5 has quotient 7 and remainder 1, the water will fill 7 jars, with **1 fl oz** left over.

— or —

We can find the remainders first, then add them. If used to fill jars that hold 5 fl oz, the first cup will have 1 fl oz left over. The second cup will have 4 fl oz left over. The third cup will have 1 fl oz left over. This is a total of 1+4+1=6 extra fluid ounces. We can use 5 of the extra fluid ounces to fill another jar, leaving 6−5=**1 fl oz**.

To measure the volume of an item that floats, push it under water without putting your fingers in the water.

There are many ways to measure the volume of an object without spilling any water. Small objects can be placed directly into a measuring cup or a graduated cylinder. First, measure the volume of the water in the measuring cup. Then, place an object in the cup so it is completely underwater. Measure the volume of the water plus the item. Subtract these two measurements to find the volume of the object.

For larger objects, fill a container with water deep enough to place the object you want to measure completely under water. Mark the height of the water before you put the object in the container. Place the object in the water, and mark the height of the water plus the object. Then, take the object out of the water and measure how much water must be added to fill the container to the higher mark. The amount of water that must be added is equal to the volume of the object.

To measure the volume of your body, fill a bathtub so that you can sink completely under water in the tub. Mark the water level in the tub. Then, get into the tub and sink

completely under water. Try not to splash or make too many waves, so that the water level stays flat. Have someone else mark the water level in the tub while you are completely under water. Get out of the tub and use a gallon jug to fill the tub to the higher water mark, keeping track of how many gallons you've added. The number of gallons you add to reach the higher water mark is your volume in gallons.

Or, you can get a good estimate of your volume in gallons by dividing your weight by 8. A gallon of water weighs a little more than 8 pounds. A typical person weighs slightly less than his or her volume in water (which is why most people float). So, dividing your weight in pounds by 8 gives your approximate volume in gallons.

The following answers represent the opinions of the authors. If you like to swim when it's freezing outside, or picnic in scorching heat, your answers may be different.

52. A good temperature for a snowball fight is near the freezing point of water, which is **0°C (32°F)**.

53. Swimming at the beach is best in hot weather. Swimming at the beach should be connected to **35°C (95°F)**.

54. To have a picnic in the park, you usually want the temperature to be comfortable. Picnic in the park should be connected to **20°C (68°F)**.

55. An evening campfire is best when the temperature is cool, but not freezing. Evening campfire should be connected to **10°C (50°F)**.

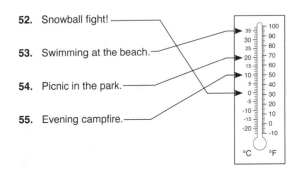

56. A penny is worth 1¢, a nickel is worth 5¢, a dime is worth 10¢, and a quarter is worth 25¢. Together, the four coins are worth 1+5+10+25=**41 cents**.

57. Ralph's four nickels and three dimes are worth a total of (4×5)+(3×10)=20+30=50 cents.
Cammie's four dimes and three nickels are worth a total of (4×10)+(3×5)=40+15=55 cents.
So, Cammie has 55−50=**5 cents** more than Ralph.

— or —

Ralph and Cammie each have 3 dimes and 3 nickels, plus one other coin. Cammie's other coin is a dime; Ralph's is a nickel. Cammie's dime is worth **5 cents** more than Ralph's nickel.

58. Fiona has $25, which is a multiple of $5. Using bills worth $5, $10, and $20, it is only possible to make multiples of $5, because the value of each bill is a multiple of $5.

If 1, 2, 3, or 4 of Fiona's bills are 1-dollar bills, the number of dollars Fiona has cannot be a multiple of 5. So, none of Fiona's bills are 1-dollar bills.

If one bill is a 20-dollar bill, then Fiona's 3 remaining bills must be worth a total of 25−20=5 dollars. No 3 bills on Beast Island are worth a total of 5 dollars. So, none of Fiona's bills are 20-dollar bills. Each of Fiona's bills must be worth $5 or $10. There are three ways to make $25 using only 5- and 10-dollar bills:

$10	$5	Total number of bills:
2	1	3
1	3	4
0	5	5

The only way that uses 4 bills has one $10 bill and three $5 bills. So, Fiona has **1** ten-dollar bill.

59. For each quarter Drew has, he also has one nickel. Together, 1 quarter and 1 nickel are worth 25+5=30 cents. Drew has 90 cents. Since 3 times 30 cents equals 90 cents, Drew has **3 quarters** and 3 nickels.

60. We can find the value of Max's money in cents, then divide by the value of one nickel to find out how many nickels Lizzie has. One dollar is worth 100 cents. Two quarters are worth 2×25=50 cents. Three dimes are worth 3×10=30 cents. So, Max has 100+50+30=180 cents. A nickel is worth 5 cents, so Lizzie has 180÷5=**36 nickels**.

— *or* —

We can find the value in nickels of Max's bill and coins. A dollar bill is worth 100 cents, and a nickel is worth 5 cents. So, Max's dollar is worth the same as 100÷5=20 nickels. A quarter is worth 5 nickels, so Max's 2 quarters are worth the same as 2×5=10 nickels. A dime is worth 2 nickels. So, Max's 3 dimes are worth the same as 3×2=6 nickels. All together, Max's money is worth the same as 20+10+6 nickels. So, Lizzie has **36 nickels**.

61. The amount of money Sally has in dimes is twice the amount she has in quarters. A quarter is worth 25 cents. So, for every quarter Sally has, she has 2×25=50 cents in dimes. Five dimes are worth 50 cents. So, for every quarter Sally has, she has 5 dimes. Another way of saying this is that for every 6 coins, Sally has 1 quarter and 5 dimes.

We can divide Sally's 18 coins into 3 stacks of 6 coins, each with 1 quarter and 5 dimes. So, Sally has 3 quarters and 15 dimes. Three quarters are worth 3×25=75 cents. 15 dimes are worth 15×10=150 cents. Sally's coins are worth 75+150=**225 cents**.

62. Grogg pays with dimes, so the price in cents for a half-pint of milk is a multiple of 10. Lizzie pays with the same number of coins, worth the same amount, but only uses quarters and pennies.

If Lizzie pays with 0 pennies, her quarters will always be worth more than Grogg's dimes. So, Lizzie cannot use 0

pennies. If Lizzie pays with exactly 1, 2, 3, or 4 pennies, she will not pay a multiple of 10 cents. So, Lizzie must use at least 5 pennies.

If Lizzie pays with 1 quarter and 5 pennies, that makes 6 coins worth 25+5=30 cents. Grogg only needs 3 dimes to make 30 cents, so this is not correct.

If Lizzie pays with 2 quarters and 5 pennies, that makes 7 coins worth 50+5=55 cents, which is not a multiple of 10, so this is not correct.

If Lizzie pays with 3 quarters and 5 pennies, that makes 8 coins worth 75+5=80 cents. Grogg needs 8 dimes to make 80 cents. So, Lizzie and Grogg can each have 8 coins worth 80 cents. If Lizzie's payment includes 10 pennies, Grogg would use at least 10 dimes, so he would pay more than 80 cents. So, **80 cents** is the smallest possible price of a half-pint of milk.

63. If Grogg has 10 coins worth 32 cents, he must have at least 2 pennies. That leaves 10−2=8 coins worth 32−2=30 cents.

The smallest amount of money Grogg can have with 8 coins if none are pennies is 8×5=40 cents (all nickels), which is too much money. So, Grogg must have more than 2 pennies.

If Grogg has 3, 4, 5, or 6 pennies, the total number of cents Grogg has with coins other than pennies will not be a multiple of 5. So, Grogg has at least 7 pennies. That leaves 10−7=3 coins worth 32−7=25 cents. Since 1 quarter is 25 cents, Grogg cannot have any quarters. Grogg cannot have 1, 2, or 3 more pennies. So, we have 3 coins that are nickels and dimes worth a total of 25 cents. This is only possible with 1 nickel and 2 dimes. So, Grogg has 7 pennies, **1** nickel, and 2 dimes.

64. To make $41 with 6 bills, at least one of the bills must be a 1-dollar bill. That leaves 6−1=5 bills worth 41−1=40 dollars. We cannot use any more 1-dollar bills, because 1, 2, 3, or 4 one-dollar bills leave an amount that is not a multiple of 5, and if we use 5 one-dollar bills, we run out of bills. So, we can make a chart showing all of the ways to make $40 using only 5, 10, and 20-dollar bills. When making a chart like this, it helps to be organized. In our chart, we found all the ways to make $40 with two $20 bills, one $20 bill, then zero $20 bills:

$20	$10	$5	Total number of bills:
2	0	0	2
1	2	0	3
1	1	2	4
1	0	4	**5**
0	4	0	4
0	3	2	**5**
0	2	4	6
0	1	6	7
0	0	8	8

We can see from the chart that there are two ways to get $40 using 5 bills. So, there are 2 ways to get $41 with 6 bills. The first way has one $20 bill, four $5 bills, and one $1 bill. The second way has three $10 bills, two $5 bills, and one $1 bill. Since Captain Kraken has more $5 bills than Ms. Q., he has **four** $5 bills.

65. The short (hour) hand is pointed at the 4, so the time is close to 4:00. The long (minute) hand points at the third tick mark. So, the time displayed on the clock is **4:03**.

66. The short (hour) hand is between the 6 and the 7, so the time is between 6:00 and 7:00. The long (minute) hand points at the 47th tick mark. So, the time displayed on the clock is **6:47**.

67. The short (hour) hand should point between the 3 and the 4. The long (minute) hand should point at the 33rd tick mark as shown below.

68. The short (hour) hand should point between the 11 and the 12. The long (minute) hand should point at the 27th tick mark as shown below.

69. The time shown on the clock is 9:19, which is 19 minutes past 9 o'clock. In 45 minutes, it will be 19+45=64 minutes past 9 o'clock. There are 60 minutes in an hour, so 64 minutes past 9 o'clock is the same as 64−60=4 minutes past 10 o'clock. We write this as **10:04**.

70. The time shown on the clock is 1:53, which is 7 minutes before 2 o'clock. Class ends at 2:15, which is 15 minutes after 2 o'clock. So, there are 7+15=**22 minutes** left in class.

71. If the movie starts at 12:25, then 1 hour after the movie begins, it will be 1:25, or 25 minutes past 1 o'clock. There are still 43 minutes left in the movie, so the movie ends at 25+43=68 minutes past 1 o'clock. There are 60 minutes in an hour, so 68 minutes past 1 o'clock is the same as 68−60=8 minutes past 2 o'clock. We write this as **2:08**, which is displayed on the clock below.

72. Adding time can be tricky. If we just add 10 hours and 10 minutes to 4:56, we get 14:66! The hour displayed on a clock is always between 1 and 12, and the minute is always between :00 and :59. Let's start by correcting the hour. 14:00 is 2 hours past 12:00, which is 2:00. So, 14:66 means 2:66.

2:66 means 66 minutes past 2:00, but since there are 60 minutes in an hour, 66 minutes past 2:00 is the same as 66−60=6 minutes past 3:00. We write this as **3:06**,

which is displayed on the clock below.

Time of day can be also given as the number of hours and minutes past midnight. In this system, one hour after 12:00 noon is 13:00 (instead of 1:00 p.m.), one minute before midnight is 23:59 (instead of 11:59 p.m.), and 30 minutes past midnight is 00:30 (instead of 12:30 am). This system is commonly called military time.

73. There are 24 hours in one day, so in three days there are 3×24=**72 hours**.

74. There are 60 minutes in one hour. Each minute has 60 seconds. So, an hour has 60×60=**3,600 seconds**.

75. Twelve hours after 5:00 p.m., it will be 5:00 <u>a.m.</u> That leaves 20−12=8 more hours to add. Seven hours after 5:00 a.m., it will be 12:00 p.m. (noon). That leaves just 8−7=1 more hour to add. One hour after 12:00 p.m. is **1:00 p.m.**

— *or* —

Instead of adding 20 hours, we can add 24 hours, then subtract 4 hours. In 24 hours, it will be the same time tomorrow as it is now. So, 24 hours after 5:00 p.m., it will be 5:00 p.m. Then, we subtract the four extra hours we added. Four hours before 5:00 p.m. is 1:00 p.m., so 20 hours after 5:00 p.m., it will be **1:00 p.m.**

76. If it is 5:00 p.m. on Wednesday, there are 12−5=7 hours left until midnight, when Thursday begins. There are 24 hours on Thursday. Then, there are 9 hours on Friday until 9:00 a.m. So, a total of 7+24+9=**40 hours** pass from 5:00 p.m. Wednesday to 9:00 a.m. Friday.

77. A millennium is 1,000 years. A decade is 10 years. So, there are 1,000÷10=**100 decades** in one millennium.

78. Every month has 28, 29, 30, or 31 days. There are 7 days in 1 week. So, 4 weeks is 4×7=28 days. This means that a 28-day month has exactly 4 weeks. A month with 28 days that begins on a Saturday will always end on a Friday. A month that begins on a Saturday and ends on a Saturday has 1 more day, so there are **29 days** in a month that begins and ends on a Saturday. This is easiest to see on a calendar.

Su	M	T	W	Th	F	Sa
						1
2	3	4	5	6	7	8
9	10	11	12	13	14	15
16	17	18	19	20	21	22
23	24	25	26	27	28	29

79. If each year had exactly 365 days, then one decade would have 365×10=3,650 days. However, a decade can have up to 3 leap years. For example, the ten years from 2000 to 2009 had leap years in 2000, 2004, and 2008. This adds three extra days to the decade, so a decade can have up to 3,650+3=**3,653 days**.

80. Since there are 60 minutes in an hour, we divide 234÷60 to find the number of hours and minutes in 234 minutes. Since 234÷60 has quotient 3 and remainder 54, there are 3 hours and 54 minutes in 234 minutes. Three hours after 2:34 p.m., it will be 5:34 p.m. From 5:34 p.m. to 6:00 p.m., 26 minutes pass. This leaves 54−26=28 minutes to add. 28 minutes after 6:00 p.m., it will be **6:28 p.m.**

— *or* —

Since there are 60 minutes in an hour, there are 60×4=240 minutes in four hours. So, 234 minutes is 6 minutes less than 4 hours. To add 234 minutes, we can add 4 hours (240 minutes), then subtract 6 minutes. Four hours after 2:34 p.m., it will be 6:34 p.m. Then, we subtract 6 minutes from 6:34 p.m. to get 6:28 p.m. So, 234 minutes after 2:34 p.m., it will be **6:28 p.m.**

81. Since Sergeant Rote's clock displays 3:17, it means that 3 hours and 17 minutes have passed since the power came back on. Since we know the real time is 6:05, we can subtract 3 hours and 17 minutes to find out what time the power came back on. Three hours before 6:05, it was 3:05. To subtract 17 minutes, we can first subtract 5 minutes to get to 3:00, then 12 more minutes to get to **2:48**.

82. This can only happen if the little monster's birthday is on December 31st, and he is asked his birthday on January 1st. Here's how it works: Let's pretend that today is January 1st, 2010. The monster turned 7 yesterday, on December 31st, 2009. So, he was 6 the day before yesterday, and he is now 7. At the very end of this year, on December 31st, 2010, the little monster will turn 8. And next year, on December 31st, 2011, the little monster turns 9.

Project: Age in Days   92

Since we don't know your birthday, or the date that you are reading this page, we don't know your exact age in days. Here are some estimates to help you figure out if the answer you got makes sense.

Your Age in Years	Approximate Age in Days
6	2,192
6½	2,374
7	2,557
7½	2,739
8	2,922
8½	3,105
9	3,287
9½	3,470
10	3,653
10½	3,835
11	4,018

83. The time it takes to microwave popcorn is best described in **seconds**. You might also use minutes and seconds. Popcorn takes about 150 seconds, or 2 min 30 sec, to microwave.

84. The weight of an orange is best described in **ounces**. Most oranges weigh between 5 and 10 ounces.

85. The volume of soda in a can is best described in **fluid ounces**. A standard soda can holds 12 fluid ounces.

86. The height of a house is best described in **feet**. A two-story house is usually between 20 and 30 feet tall.

87. The capacity of a trash can is best described in **gallons**. A kitchen trash can has a capacity of about 10 gallons. A large, wheeled trash can holds more than 50 gallons.

88. The age of a large tree is best described in **years**. While most full-grown trees in your neighborhood are probably between 10 and 100 years old, some trees can live to be extremely old. The oldest known trees have lived to be over 4,000 years old and are the oldest living things on the planet!

89. The width of this book is best described in **centimeters**. This book is between 21 and 22 centimeters wide when closed.

90. The weight of a jumbo jet is best described in **tons**. You could use pounds, but a jumbo jet can weigh close to a million pounds! This is about 500 tons.

91. An adult caterpillar is about 5 **centimeters (cm)** long.

92. My cup contains 11 **fluid ounces (fl oz)** of juice. You could also have correctly written 11 **ounces (oz)**, but juice is usually measured in fluid ounces.

93. I can ride my bike 10 **miles (mi)** in one hour.

94. An eyedropper holds about 2 **milliliters (mL)** of liquid. You may have also correctly written 2 **grams (g)**, but liquid is usually measured by volume.

95. The SUV weighs about 3,500 **pounds (lb)**.

96. The height of the kitchen counter is 34 **inches (in)**.

97. The football game on television was 3 **hours (hr)** long.

98. Together, a nickel and two pennies weigh about 10 **grams (g)**.

99. An adult elephant weighs about 5 **tons**.

100. It takes about 3 **seconds (sec)** to read this sentence.

101. A toilet uses about 10 **liters (L)** of water per flush.

102. A dollar bill is **a little more than 6** inches long. If you guessed between 4 and 8 inches, you were pretty close.

103. By plane, the flight from New York to Los Angeles is about 2,500 miles. The driving distance is closer to 2,800 miles. Any guess **between 2,000 and 3,000** miles is really good. If your guess was between 1,500 and 4,000 miles, you were pretty close.

104. A typical box of cereal usually weighs **between 10 and 20 ounces**. Check your cupboards to see how close your guess was.

105. The sentence is **a little less than 8 centimeters** long. If your guess was between 5 and 15 centimeters, you were pretty close.

106. A quarter weighs **between 5 and 6 grams**. If your guess was between 5 and 10 grams, you were pretty close. You may remember that we gave you the weight of a nickel on page 79.

107. Shampoo bottles come in many sizes, but a typical bottle holds **between 10 and 20 fluid ounces**. Shampoo bottles can be larger or smaller. Check your shampoo bottle to see how close your guess was.

108. A gallon of milk weighs **between 8 and 9 pounds**. If your guess was between 5 and 12 pounds, you were pretty close.

109. Kitchen sinks come in all shapes and sizes, but most kitchen sinks hold **between 10 and 20 gallons**. You can estimate the volume of your kitchen sink by measuring its length, width, and depth in inches. Multiply these three numbers, then divide by 230 to get a pretty close estimate of the sink's capacity in gallons.

110. Five kilometers is about the same distance as 3 **miles**.

111. One liter is a little more than 1 **quart**.

112. One kilogram is a little more than 2 **pounds**.

113. Two and a half centimeters is about the same length as 1 **inch**.

114. One meter is a little longer than 1 **yard**.

MEASUREMENT
Challenge Problems   96-98

115. There are 60 minutes in an hour. Since each quarter allows Pat to play for 5 minutes, it will take 60÷5=12 quarters to play for one hour. There are 4 quarters in a dollar, so it will take 12÷4=**3 dollars** worth of quarters for Pat to play the video game for an hour.

116. We can convert Dave's marathon time into minutes, find half of Dave's time in minutes, then convert back to hours and minutes. There are 60 minutes in an hour, so 5 hours and 18 minutes equals 5×60+18=318 minutes. Half of 318 is 159, so Patrick's best marathon time is 159 minutes. Two hours equals 2×60=120 minutes, so 159 minutes is 159−120=39 minutes more than 2 hours. So, Patrick's best marathon time is **2 hr 39 min**.

— *or* —

We can find half of 5 hours, and add it to half of 18 minutes. Half of 5 hours is 2 hours and 30 minutes. Half of 18 minutes is 9 minutes. So, half of 5 hours and 18 minutes is 2 hours + 30 minutes + 9 minutes, which equals **2 hr 39 min**.

117. We can convert both measurements to fluid ounces and subtract to find out how much water is in the jug now. There are 8 fluid ounces in 1 cup, so 3 cups and 7 fluid ounces equals 3×8+7=31 fluid ounces. 7 cups

and 3 fluid ounces equals 7×8+3=59 fluid ounces. We subtract 31 fluid ounces from 59 fluid ounces to get 59−31=28 fluid ounces.

Then, to convert 28 fluid ounces to cups and fluid ounces, we divide 28 by 8. Since 28÷8 has quotient 3 and remainder 4, there are 3 whole cups and 4 extra ounces in 28 fluid ounces. We can write this as a mixed measure in cups and fluid ounces: **3 cups 4 fl oz**.

— *or* —

To subtract 3 cups 7 fluid ounces from 7 cups 3 fluid ounces, we can subtract cups and ounces separately. Unfortunately, there are more fluid ounces in the number we are subtracting.

7 cups	3 fl oz
− 3 cups	7 fl oz

So, we convert one of the cups from 7 cups 3 ounces into fluid ounces. This gives us 6 cups 11 ounces. Then, we can subtract.

6 cups	11 fl oz
− 3 cups	7 fl oz
3 cups	4 fl oz

So, there are **3 cups 4 fl oz** in the jug.

118. In 8 feet, there are 8×12=96 inches. We need to find the number of cups that equal 96 ounces. There are 8 ounces in 1 cup. Since 96÷8=12, there are 12 cups in 96 ounces. So, the number of ounces in **12 cups** equals the number of inches in 8 feet.

— *or* —

In 8 feet, there are 8×12 inches. We need to find a number of cups that will give us 8×12 ounces. Since there are 8 ounces in one cup, **12 cups** equal 8×12 ounces.

119. Gary's pack weighs as much as four Garys. Together, Gary and his pack weigh as much as 5 Garys. We know the weight of Gary and his pack is 1 lb 4 oz. There are 16 ounces in 1 pound, so 1 lb 4 oz equals 16+4=20 ounces. If the weight of 5 Garys is 20 ounces, than Gary weighs 20÷5=**4 ounces**.

120. We follow the instructions to convert 40°C to degrees Fahrenheit. First, we multiply by 9 to get 40×9=360. We divide by 5 to get 360÷5=72. Finally, we add 32 to get 72+32=104. So, 40°C equals **104°F**.

121. To find how many seconds Grogg's snail takes to slime 5 yards, we convert 5 yards to inches. 1 yard equals 3 feet, and there are 12 inches in 1 foot. So, 1 yard equals 3×12=36 inches. This means that 5 yards equals 36×5=180 inches. Grogg's snail slimes 1 inch every second. So, to slime 180 inches, Grogg's snail takes 180 seconds. We need to find how many minutes this is. Since there are 60 seconds in 1 minute, there are 180÷60=3 minutes in 180 seconds. Grogg's snail takes **3 minutes** to slime 5 yards.

122. There are 1,000 grams in one kilogram. So, 400 pennies weigh 1,000 grams. We can divide 400 pennies into 10 stacks so that each stack of 400÷10=40 pennies weighs 1,000÷10=100 grams. So, 40 pennies weigh 100 grams. Yerg's penny collection weighs 700 grams, which is the same as seven 100-gram stacks. There are 40 pennies in each 100-gram stack. So, Yerg has 7×40=**280** pennies.

123. If the power goes off at 3:00 and stays off for 45 minutes, Klurg's analog clock will begin running again at 3:45, but will read 3:00. So, the time displayed on Klurg's analog clock will be 45 minutes earlier than the actual time. At 7:00, Klurg's analog clock will display 45 minutes before 7:00, or **6:15** as shown below.

Klurg's digital clock reads 12:00 when the power comes back on at 3:45. So, the time displayed on Klurg's digital clock at 7:00 is 3 hours and 45 minutes earlier than the actual time. Three hours before 7:00 is 4:00, and 45 minutes before 4:00 is **3:15** as shown below.

124. When the power came back on, the digital clock reset to 12:00. So, the time on the digital clock tells us how long ago the power came back on. The time on the digital clock is 4:05. So, the power came back on 4 hours and 5 minutes ago.

When the power came back on, the analog clock displayed the exact time that the power went out. It has been 4 hours and 5 minutes since the power came back on. So, to find the time the power went out, we subtract 4 hours 5 minutes from the time on the analog clock, which is 6:10. Four hours before 6:10 is 2:10, and 5 minutes before 2:10 is **2:05**.

Notice that to find the time the power went out, we can subtract the time on the digital clock from the time on the analog clock. There is no way of knowing the current time using only the times displayed on the two clocks.

125. Since 4+2+1=7, we can balance the scale by placing a 4-gram weight, a 2-gram weight, and a 1-gram weight on the right side as shown.

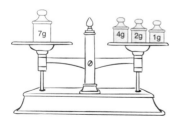

126. Since 8+4+1=13, we can balance the scale by placing an 8-gram weight, a 4-gram weight, and a 1-gram weight on the right side as shown.

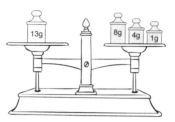

127. Professor Grok can balance a scale that holds a coin that weighs up to 15 grams. The equations below show how 1-, 2-, 4-, and 8-gram weights could be placed to balance a scale holding each weight from 1 to 15 grams.

1=1	6=4+2	11=8+2+1
2=2	7=4+2+1	12=8+4
3=2+1	8=8	
4=4	9=8+1	13=8+4+1
5=4+1	10=8+2	14=8+4+2
		15=8+4+2+1

We do not have enough weight to balance a scale that holds a 16-gram coin. So, **16 grams** is the smallest possible weight of Captain Kraken's coin.

128. To balance the scale, we can place a 27-gram weight on the right side and a 9-gram weight on the left side. We see that the left side has 18+9=27 grams, and the right side has 27 grams. So, the scale is balanced.

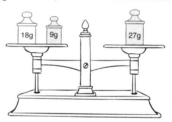

129. To balance the scale, we can place a 27-gram and a 9-gram weight on the right side, and a 1-gram weight on the left side. We see that the left side has 35+1=36 grams, and the right side has 27+9=36 grams. So, the scale is balanced.

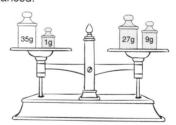

130. Professor Grok can balance a scale that holds a tribble that weighs up to 40 grams. The equations below show how 1-, 3-, 9-, and 27-gram weights could be placed to balance a scale that holds a tribble whose weight in grams is given in bold.

1=1	**15**+9+3=27	**28**=27+1
2+1=3	**16**+9+3=27+1	**29**+1=27+3
3=3	**17**+9+1=27	**30**=27+3
4=3+1	**18**+9=27	**31**=27+3+1
5+3+1=9	**19**+9=27+1	**32**+3+1=27+9
6+3=9	**20**+9+1=27+3	**33**+3=27+9
7+3=9+1	**21**+9=27+3	**34**+3=27+9+1
8+1=9	**22**+9=27+3+1	**35**+1=27+9
9=9	**23**+3+1=27	**36**=27+9
10=9+1	**24**+3=27	**37**=27+9+1
11+1=9+3 **25**+3=27+1		**38**+1=27+9+3
12=9+3 **26**+1=27		**39**=27+9+3
13=9+3+1 **27**=27		**40**=27+9+3+1
14+9+3+1=27		

Notice that a scale holding a tribble that weighs from 1 to 13 grams can be balanced using only the three smallest weights.

To balance a scale that holds a 14-gram tribble, we start by putting the 27-gram weight on the right. That gives us 27−14=13 more grams on the right side of the scale than we have on the left. So, the problem is the same as balancing a scale that holds a 13-gram tribble using the three smaller weights. This strategy works for each weight from 14 to 40 grams.

For example, we can balance a scale that holds a 23-gram tribble by first placing the 27-gram weight on the opposite side. Then, the problem is the same as using the three smaller weights to balance a scale that holds 27−23=4 grams. Or, to balance a scale that holds a 34-gram tribble, we start by placing the 27-gram weight on the opposite side. Then, the problem is the same as using the three smaller weights to balance a scale that holds 34−27=7 grams. We can use this strategy to balance a scale that holds a trible weighing up to 27+9+3+1=40 grams.

We do not have enough weight to balance a scale that holds a 41-gram tribble. So, **41 grams** is the smallest number of grams Grogg's tribble could weigh.

MEASUREMENT
Investigations 99

You may have come up with a different way to complete each task in Problems 131-133. Alternate solutions are always encouraged. If you know a great way to solve one of the problems below, write us at info@BeastAcademy.com. Maybe we'll include your solution in the next printing of this book!

131. A ruler can be used as a balance. Place a 12-inch ruler on an object like a pencil so that it balances. It should balance at or near the 6-inch mark. Then, place a stack of 4 quarters on one side of the ruler, and a stack of 10 dimes on the other. The center of the stacks should be the same distance away from the balance point (called the fulcrum). The side that is heavier will go down. In this case, ten dimes should weigh exactly the same as four quarters (22.68 grams).

132. When lightning strikes, it makes a loud boom called thunder. You may have noticed that you often see the flash of lightning before you hear the thunder. This is because sound travels much slower than light. Light travels so fast that you see the flash of lightning at almost the exact time it strikes. Light travels about 186,000 miles per second! At this speed, light reflected off of the moon takes only about 2 seconds to reach your eyes.

Sound is fast too, but not nearly as fast as light. Sound travels at about 760 miles per hour, which is a little more than 1,100 feet per second. So, to find how many feet away the lightning strikes, time the number of seconds between the moment you see the flash and the moment you hear the thunder and multiply by 1,100.

For example, if you see a flash of lightning and 4 seconds pass before you hear the thunder, the strike was about 4×1,100=4,400 feet away. If you count 5 seconds between the time you see lightning strike and the time you hear the thunder, the strike was about 5,500 feet away, or about 1 mile.

133. You have probably seen an hourglass used to measure time. An hourglass measures a specific amount of time by allowing sand to pass through a small hole from the top of the hourglass to the bottom. We can use the same idea to measure time with a milk jug. Poke a small hole in the bottom of an empty milk jug. Plug the hole with your finger and fill the jug with water.

Unplug the hole and time how long it takes for the water to drain. It may help to mark the water level every 15 seconds or so to help you figure out how quickly the water drains. Experiment until you have a milk jug that takes 5 minutes to drain. You can change the size of the hole, or the amount of water you put in the jug.

— *or* —

Hang the milk jug from a string to create a pendulum as described earlier in the chapter. Pendulums have been used to keep time for centuries! You can change the length of the pendulum to get a specific period length. For example, a pendulum that is slightly less than 4 meters long will have a 4-second period. So, in 1 minute, it will make 15 full out-and-back swings. In 5 minutes, it will make 75 swings.

For additional books, printables, and more, visit
www.BeastAcademy.com